Fundamentals of Radio Frequency Identification

Fundamentals of Radio Frequency Identification

Cillian Shelby

STATES
ACADEMIC PRESS
www.statesacademicpress.com

Fundamentals of Radio Frequency Identification
Cillian Shelby
ISBN: 978-1-63989-232-7 (Hardback)

STATES
ACADEMIC PRESS

Published by States Academic Press,
109 South 5th Street,
Brooklyn, NY 11249, USA

Cataloging-in-Publication Data

Fundamentals of radio frequency identification / Cillian Shelby.
p. cm.
Includes bibliographical references and index.
ISBN 978-1-63989-232-7
1. Radio frequency identification systems. 2. Identification--Equipment and supplies.
3. Near-field communication. 4. Radio telemetry. I. Shelby, Cillian.
TK6570.I34 F86 2021
621.384 192--dc23

Table of Contents

Preface

This book has been written, keeping in view that students want more practical information. Thus, my aim has been to make it as comprehensive as possible for the readers. I would like to extend my thanks to my family and co-workers for their knowledge, support and encouragement all along.

The use of electromagnetic fields to detect and track tags attached to various objects is known as radio frequency identification. It has three components-radio transponder, a radio transmitter and a receiver. Whenever a tag is triggered by an electromagnetic pulse, it transmits digital data back to the receiver. This is the underlying principle on which this domain operates. These devices can be attached to animals, humans, clothes, cash and possessions. These can be applied in diverse areas such as tracking of animals and persons, access management, airport baggage tracking logistics, machine readable travel documents, tracking and billing processes, tracking of goods, contactless payment and toll collection, smartdust, timing sports events, etc. The book aims to shed light on some of the unexplored topics of this field. Most of the topics introduced in this book cover new techniques and applications of radio frequency identification. Those in search of information to further their knowledge will be greatly assisted by this book.

A brief description of the chapters is provided below for further understanding:

Chapter - Introduction

The oscillation rate of an alternating electric current, electromagnetic field and mechanical system in the frequency range of 20kHz to 300GHz is termed as radio frequency. Radio-frequency identification (RFID) is the use of electromagnetic fields to read and capture information stored on a tag attached to an object. This chapter sheds light on radio frequency identification for an in-depth understanding of the subject.

Chapter - Radio Frequency Identification Tags

An electronic tag that exchanges data with an RFID reader through radio waves is referred to as a radio frequency identification tag. It includes active tags, passive tags, clipped tag, bag tag, organic RFID tags, etc. This chapter delves into the subject of radio frequency identification tags to provide a thorough understanding of it.

Chapter - Radio Frequency Identification Middleware

The RFID middleware is a set of components which aims to manage RFID readers, deals with RFID-events and data, and is connected to end-user applications. The middleware is located between readers and business applications. All the aspects related to RFID middleware have been carefully analyzed in this chapter.

Chapter - Issues and Challenges

There are various security, privacy, ethical and technology related problems with RFID. It includes RFID tag collision, RFID reader collision, zombie RFID tags, etc. This chapter has been carefully written to provide an easy understanding of these issues and challenges in radio frequency identification.

Chapter - Applications

Radio frequency identification has a wide-range of applications such as in logistics and supply chain visibility, item level inventory tracking, attendee tracking, materials management, access control, IT asset tracking, library systems, interactive marketing, etc. This chapter discusses these applications of radio frequency identification in detail.

Cillian Shelby

1

Introduction

The oscillation rate of an alternating electric current, electromagnetic field and mechanical system in the frequency range of 20kHz to 300GHz is termed as radio frequency. Radio-frequency identification (RFID) is the use of electromagnetic fields to read and capture information stored on a tag attached to an object. This chapter sheds light on radio frequency identification for an in-depth understanding of the subject.

RADIO FREQUENCY

Radio frequency (RF) is a measurement representing the oscillation rate of electromagnetic radiation spectrum, or electromagnetic radio waves, from frequencies ranging from 300 GHz to as low as 9 kHz. With the use of antennas and transmitters, an RF field can be used for various types of wireless broadcasting and communications.

Working

Radio frequency is measured in units called hertz, which represent the number of cycles per second when a radio wave is transmitted. One hertz equals one cycle per second; radio waves range from thousands (kilohertz) to millions (megahertz) to billions (gigahertz) of cycles per second. Microwaves are a type of radio wave with higher frequencies. Radio frequencies are not visible to the human eye.

In a radio wave, the wavelength is inversely proportional to the frequency. If f is the frequency in megahertz and s is the wavelength in meters, then:

$$s = 300/f$$

As the frequency is increased beyond that of the RF spectrum, electromagnetic energy takes the form of infrared (IR), visible, ultraviolet, X-rays and gamma rays.

RF Technology

Many types of wireless devices make use of RF fields. Cordless and cellphones, radio and television broadcast stations, Wi-Fi and Bluetooth, satellite communications systems, and two-way radios all operate in the RF spectrum. In addition, other appliances outside of communications, including microwave ovens and garage-door openers, operate at radio frequencies. Some wireless devices, like TV remote controls, some cordless computer keyboards and computer mice, operate at IR frequencies, which have shorter electromagnetic wavelengths.

The RF spectrum is divided into several ranges, or bands. With the exception of the lowest-frequency segment, each band represents an increase of frequency corresponding to an order of magnitude (power of 10). The following table depicts the eight bands in the RF spectrum, showing frequency and bandwidth ranges. The super high frequency (SHF) and extremely high frequency (EHF) bands are often referred to as the microwave spectrum.

Table: Radio Frequency Spectrum bands.

Designation	Abbreviation	Frequencies	Free-space Wavelengths
Very Low Frequency	VLF	9kHz to 30 kHz	33 km to 10km
Low frequency	LF	30kHz to 300 kHz	10km to 1km
Medium Frequency	MF	300kHz to 3MHz	1 km to 100m
High Frequency	HF	3MHz to 30 MHz	100 m to 10m
Very High Frequency	VHF	30MHz to 30 MHz	10 m to 1 m
Ultrahigh Frequency	UHF	30MHz to 3GHz	1 m to 100mm
Super-high Frequency	SHF	3GHz to 30 GHz	100mm to 10mm
Extremely high Frequency	EHF	30GHz to 300 GHz	10mm to 1mm

In the United States, radio frequencies are divided into licensed and unlicensed bands. The Federal Communications Commission issues licenses that permit a commercial entity to have exclusive use of a frequency band in a given location. Unlicensed frequencies are free for public use, but remain a shared medium.

RADIO FREQUENCY IDENTIFICATION

Radio Frequency Identification (RFID) technology, as the name suggests is a method of wirelessly transferring data between transceiver and transponder where the transponder (tag) is attached to a variety of objects and used to automatically identify each object across a variety of industrial purposes. The fundamentals of RFID technology are based on the principles surrounding electromagnetic energy, originating with Michael Faraday's explanation of light and radio waves dating back to 1846. This technological approach has the ability to identify virtually any object in any location in real time. Tags, may be attached to any item either living or in animate have the ability to manipulate propagating radio waves in such a fashion as to enable them to carry data, which may be customised to the item (item level traceability), an approach which is core to the functioning of these systems. This provides us with the core competency of RFID which is its ability to provide uniqueness to each individual item and make it identifiable in real time making RFID technology one of the most popular auto ID technologies in use to date.

A basic RFID system consists of three main components : (1) Tag, (2) Reader and (3) Antenna which are collectively responsible for the generation of the radio wave. The tag (transponders) consists of an integrated circuit, an antenna, and may or may not have an onboard power source. They are classified as active (i.e., contain an on-board power source to cater for all electrical power needs); passive (i.e., no internal power source and rely on the interrogation wave from the reader for all power requirements), or semi-passive/semi-active (i.e., contain their own power source, yet rely on the interrogation wave for coupling, while the on-board source is used to power the tag circuitry). The RFID market has gathered tremendous traction in the area of environmental temperature monitoring. Sensor tags are becoming more and more commonplace across many industries and common users of this type of technology include argi-food pharmaceutical and defense supply chains.

The reader (transceiver), may be either stationary or portable and are responsible for transmitting an electromagnetic wave via in built or external antenas which communicates (couples) with the transponders via radio signal propogation using a fitted antenna. The reader is usually connected to a PC or contains an internal PC, in handheld versions, which provides a user interface to the operator. The third and final component is the antenna which emits a radio wave and as they are the primary medium of communication between tag and reader their importance cannot be over emphasised. The communication process between tag and reader is commonly referred to as coupling.

RFID technology may be grouped further into a frequency based caterogisation i.e. Low frequency (LF), High Frequency (HF) and Ultra High frequency (UHF). With each frequency variation come differing operational characteristics thus making the selection of a particular system dictated mainly by the desired end use (application environment) of the system. For many years RFID technology has been fine tuned to the extent that each of the individual frequencies have been tailored to a number of particular commercial applications. It is important to consider frequency when selecting RFID systems for a particular application as many unsucessful applications are based on incorrect frequency selection leading to an inability to acquire the true benefits of the technology.

Working of RFID

Irrespective of tag type, classification, antenna configuration, orientation or operating frequency RFID systems require the establishment of a wireless link between tag and reader to successfully communicate (couple) without which the system cannot function. This wireless link is used to carry encoded data signals thereby forming a medium for data transfer across non-physically linked objects. As with all wireless communication systems the power of the signal in relation to the frequency and communication protocol are bound by a global governing authority. This is to ensure system standardisation and interoperability as radio waves form the primary method of communication between tag and reader their importance cannot be over emphasised. An important metric in the classification of RFID systems is their method of coupling. A broad generalisation differentiates RFID systems based on near field or far field communication protocols and below is a brief introduction to both principles.

In Low and High frequency systems the antenna are structured in such a way that the propagation component of the propagating wave is small or even non-existent and these systems operate in the near field via a process called load modulation. Near field operation relies on magnetic coupling and is the chosen protocol for LF and HF systems. A current flowing through a coil will lead to the formation of a magnetic field in the immediate surrounds. The magnetic field is high in the immediate environment of the transmitting coil yet decays rapidly (inverse cubed proportion) with distance.

Backscatter coupling occurs in the far field region and the short wavelengths of these frequencies enable the construction of transponder antenna significantly smaller in dimension than lower frequency systems facilitating RFID identification down to item level. In this communication link the field effectively breaks away from the antenna and as a result communication occurs in the far field. In this type of coupling the reactive component is small and the propagationary component dominates. It is achieved by modulating the radar cross section of the tag antenna. This is based on how close the tag is to the reader antenna and on the operating frequency of the RFID system. Current flows on a transceiver antenna resulting in an electromagnetic wave. This propagating wave is picked up by a tag antenna lying within range, inducing a voltage, a portion of which is used to power interal cuirtutry and the remaining portion used to generate the back scattered signal.

Functions of RFID

Functions of a radio frequency identification (RFID) tagging system include internal control of fixed assets, internal control of expensed assets and theft prevention. The RFID system involves a tag and a reader. When scanned, the antenna within the tag picks up the radio wave and sends a response back to the reader.

Physical Inventories

During a physical inventory/audit, the auditor scans the RFID tag on each asset. This triggers a transmission that "marks off" the assets in the property record database. The auditor may then print the list of assets and see which were not located.

Internal Control of Expensed Assets

Companies may choose to use RFID tags on smaller assets as well, since a good property control system includes all corporate assets, especially those that might get stolen or lost. The company should use a separate system or file for expensed items, since it is the capitalized assets that generate periodic depreciation expense and must be reconciled with the financial statements. Internal control should not be contingent on the cost of an item or capitalization policies.

Theft Prevention

Companies with a high volume of theft or in an industry with highly classified information may choose to install RFID scanners at the door. The scanner would alert the receptionist or security guard if an employee tried to leave with a tagged item.

RFID vs. Barcode Tagging

The primary disadvantage of barcode tagging is that the tags can get dirty and become hard to read. The RFID tag combats this disadvantage because the reader needs to be only within a few feet of the tag; it doesn't require direct line of sight for reading. It also cuts down the time required to complete a physical inventory because of how it works.

The primary disadvantage of RFID tags is the cost; they are much more expensive, partly because it requires system integration and an integrator. However, the time saved (including hourly pay for the inventory analyst) may outweigh the initial cost.

Advantages and Disadvantages of Radio Frequency Identification

Scanning Range

An RFID reader can scan a tag as long as it is within frequency range. It does not have any line-of-sight limitations. Alternative ID solutions, such as barcoding, require the reader to be close to the barcode before it can "see" it to scan it. RFID systems can automatically pick up tag IDs from a distance and, in some cases, through obstacles between the tag and the reader.

RFID Capabilities

RFID systems can scan multiple items simultaneously. For example, you could scan incoming goods in your warehouse in the box, allowing you to check all contents at once without having to run individual barcode scans on each item. Other ID systems typically have a single or limited identifier for each object -- RFID tags can contain more information. Some are also read-write, allowing you to add or change data. You can implant tags into objects or use plastic coverings to protect them. This makes them more robust than some other ID tags. For example, barcodes must sit on the exterior of objects, making them prone to damage that may make them unreadable.

Speed and Convenience

RFID readers can scan tags in milliseconds and work automatically. Optical scanning systems may need manual operation and may work less quickly, since the operator has to align the reader and code exactly to scan it successfully. The speed of operation also has convenience benefits in services such as cashless payments. For example, some festivals, venues and theme parks allow visitors to load cash onto RFID-tagged wristbands so that they can tap a reader to pay. They do not have to carry wallets with them and may spend less time waiting in line.

RFID Costs

Although RFID technology has been around since the 1970s, its initial high costs restricted usage to larger businesses, many of whom developed proprietary systems. Although costs are falling, RFID systems are still typically more expensive to set up and use than alternative systems such as optical scanning. However, RFID systems bring their own cost benefits, such as reduced labor costs and improved efficiency.

Scanning Issues

Despite their reliability, RFID systems can still have problems. Although readers can scan through most non-metallic materials, they have problems with metal and water. The fact that you can scan multiple objects in a range is a benefit, but also comes with possible issues that can cause malfunctions. Tag collision may occur if a reader picks up signals from multiple tags at the same time. Reader collision may be an issue if two readers interfere with each other's signals.

RFID Security and Privacy Concerns

RFID also brings up some security issues. Unauthorized devices may be able to read and even change data on tags without the knowledge of the person who owns the object. Side-channel attacks can pick up RFID data as it passes from a tag to a reader, which could give the attacker access to passwords or information that should be secure. Some states have privacy statutes to restrict activities that might use RFID technology to collect personal information.

Applications of Radio Frequency Identification

Major applications of RFID are given below:

- Healthcare Applications: RFID applications in healthcare could save important resources that can further contribute to better patient care. RFID applications could reduce the number of errors by tagging medical objects in the healthcare setting such as patients' files and medical equipment tracking in a timely manner. RFID further improves the situation for patients' care by integrating medical objects involved throughout the patients care. RFID based timely information about the location of objects would increase the efficiency and effectiveness of paramedical staff leading to improved patients; experience.

- Baggage Applications: Airline industries, package and delivery service lose a lot of money on lost or late delivery of baggage/packages. Handling large amount of packages from many places to various destinations on different routes can be very complex. In this scenario RFID application provide best resource management, effective operation and efficient transfer of packages. RFID helps to identify the packages, and provide records that can advise the industry on possible areas that may require some improvements. It also keeps customers informed about their packages.

- Toll Road Applications: RFID applications make the toll collection/charging better with improved traffic flow, as cars/vehicles cannot pass through toll stations without stopping for payment. RFID is used to automatically identify the account holder and make faster transactions. This application helps to keep good traffic flow and to identify traffic patterns using data mining techniques that can inform the administration or decision support systems. For example, the information can be used to report the traffic conditions or to extend and develop future policies.

- Asset Tracking and Locating Objects: RFID can be used to prevent misplacement of items, or to locate items. An asset is tagged with RFID chip for its physical verification. A database is used to keep track of item movements.

- Libraries of RFID Labels: RFID can be used in library for management of the books. For this management RFID use many components like tag, reader, self-check-out/in, book drop reader, middleware etc. With the help of these components it manages the process of borrowing and return the book. RFID remembers to be already borrowed books while borrowing the book and already returned book while returning the book.

- Animal Identification: This is one of earliest RFID application. RFID tag can be injected to remain under skin of animal. This process is less painful and there is no identification mark with the help of which tag can be removing or modify. RFID chip inside tag is “Read-only” so data cannot be modified. This chip contain many information like date of birth, last vaccination done, any medical history and distinguishing features about the animal.

- Anti-Theft System: Any item can be protected by using RFID anti-theft tag is attached by a strong string or a plastic band to the item. If anyone walks to the exit with this item, RFID door antennas placed near exit will detect the presence of tag and sound an alarm.

- Waste Management: RFID can be used for waste management also. RFID tag is attached to each waste bin and every garbage truck has RFID reader attached to it. When waste bin is emptied into truck then reader read tag and transmits data to truck driver's cabin wirelessly. At end of route data is transmitted to central server. This data include waste bin number, collected at what time, who was waste collector.

- National Identification: National identification has been a biggest problem for all the countries. For identification RFID technology can be used. A user has only a single card with embedded RFID chip. This RFID tag number then points to an online database which is accessed by multiple agencies. A single ID card needs to be issued for identification.

The RFID tag can be attached to animals, plants and in particular human body. The technology is capable of preventing medical accidents in the health industry. RFID tag system is able to obtain and store blood pressure and body temperature. RFID sensors have found application in health care. RFID sensors have been used to monitor the heart-rates of cardiac patients, to identify patients for surgery, and to monitor the life of dental retainers. Lin et al. propose a RFID-based information management system for wirelessly monitoring the missile assembly process. Torrent et al. use a combination of global positioning systems and RFID-tags to monitor the components arriving at a construction site. Other applications include an efficient paper roll management system and improved asset management and accountability.

CHARACTERIZATION OF THE IDENTIFICATION PROCESS IN RFID SYSTEMS

Radio Frequency Identification (RFID) is increasingly being used to identify and track objects in supply chains, manufacturing process, product traceability, etc. These environments are characterized by a large number of items which commonly flow in conveyor belts, pallets and lorries, entering and leaving logistic installations. In these scenarios the RFID systems are installed as follows: one or more readers are placed in a strategic place, creating checking areas. The tags, attached to items, enter and leave the checking areas (traffic flow). The goal of RFID in these applications is to guarantee the communication with the tags as quickly and reliably as possible, ensuring that all tags are identified before they leave the checking areas.

One of the main problems related to RFID in these applications is that both readers and tags share the RF spectrum. Hence, when two or more tags/readers transmit simultaneously a collision occurs. The collisions diminish the system performance, producing a delay in the identification, and may cause tags leaving the workspace unidentified. The parameter that measures the rate of unidentified tags is the Tag Loss Ratio (TLR).

Depending on the application, even a TLR = 10^{-3} may be disastrous and cause thousands of items lost per day. The collisions resolution on RFID has been extensively studied during the last years, for active and passive RFID systems. In active RFID, the collision resolution is not only mandatory to reduce the identification time, but also to decrease the tag energy consumption in order to maximize the batteries lifetime. In this case, tag hardware permits to put forward sophisticated anti-collision mechanisms. Nevertheless, this complex hardware also entails high-cost-devices and, in the end, the tag price becomes the dominant factor in the final deployment.

On the other hand, in passive RFID, the extreme simplicity of the tags is a hard constraint for the design of new collision resolution methods. However, the low-cost-price of passive tags is its most attractive characteristic which permits to think about a massive adoption in a near future. Several proposals have been conducted during the last years with the aim at minimizing the collision problems in passive RFID systems, suggesting new anti-collision protocols that, a priori, outperform the current standards. Most of these studies have been addressed assuming static scenarios, that is, populations of tags that enter the workspace and stay there until all of them are identified, computing throughput (rate of identified tags per time unit), and the mean identification time. Although these results provide us insight into the performance of the algorithms, do not help to discern the conditions and phenomena which render to have uncontrolled tags (tags which are not identified when they leave the workspace) in dynamic scenarios, i.e., scenarios where tags are entering and leaving the workspace.

Identification Process

Passive RFID technology has been inevitably selected in the majority of the industrial systems with a large number of identification objects. Several reasons can be adduced: The main one is the extremely low-cost of the tags (prices below 0.10 €), as well as lack of maintenance for the tags, reusability, easy installation, etc. Passive RFID systems are installed in industrial environments to collect, automatically and transparently, the information regarding the items that enter and leave the workspace. The information is stored and managed by means of specialized middleware and software. Thus, updated information can be managed in real-time, decreasing the time to recognize, find, locate and manage items, therefore, improving facilities. Besides, RFID makes product traceability possible, which is an important issue in some industrial sectors.

As stated in the introduction, passive RFID system consists of one or more readers or interrogators placed in strategic zones and a potentially large population of cheap and small devices called tags or transponders. The readers transmit electromagnetic waves continuously, creating checking areas. Tags enter and leave the checking areas.

Passive tags are composed by an antenna, a simple electronic circuitry and a minimum amount of memory where it stores some information about the object (e.g. standard codes, history of transactions, expiration date). Since passive tags do not incorporate their own battery, they obtain the energy from the electromagnetic waves emitted by the reader (backscatter procedure). This energy activates the electronic circuitry of the tag, which delivers a signal response with its carried data. Nevertheless, the simplicity of the tags limits their operative range, varying from some centimetres to a pair of meters. The uplink and downlink communication between the reader and the tags share the RF spectrum. When several tags are in the coverage area at the same time, collisions may occur as a result of simultaneous transmissions. Hence, Medium Access Control (MAC) protocols are needed to handle/avoid collisions, but the extreme simplicity of the tags constraints the design of suitable anti-collision protocols. Complex or sophisticated behaviour can only exist in the reader system.

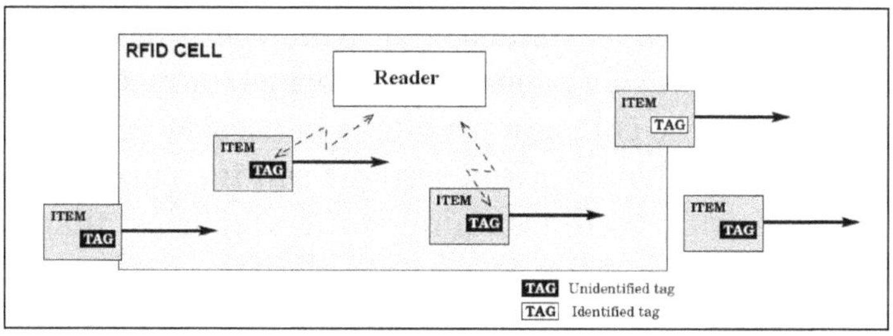

Passive RFID scenario.

During the last decade some standards have been proposed, not only for clarifying the specifications of hardware on passive tags and readers, but also for setting the anti-collision procedure to handle/avoid collisions. Since 2005, the standard EPCglobal Class-1 Gen-2 for passive RFID has been the most extended and adopted by the manufacturers. However, it has been outperformed by new anti-collision algorithm proposals. Most of these studies have been addressed assuming static scenarios, that is, blocks of tags entering the checking area and staying there until all of them are successful identified. Disappointedly, actual RFID systems may involve a flow of tags entering and leaving the workspace. In these scenarios, the collision resolution is not only mandatory to reduce the identification time and to improve the system throughput, but also for minimizing the number of unidentified tags that leave the workspace, that is, the TLR.

The Collision Problem: Anti-collision Mechanisms for RFID

Several multiple-access techniques and anti-collision protocols have been developed with the aim of minimizing colliding signals. The anti-collision procedures are focused on the Physical and MAC layer.

Anti-collision Protocols at Physical Layer

At physical layer, FDMA, TDMA, SDMA, CDMA and CSMA have been the alternatives most studied. Albeit they cannot be used directly in RFID because the following problems:

- Frequency Division Multiple Access (FDMA): The channel is divided into different subchannels and the users are allocated different carrier frequencies. In

RFID systems, this technique adds a cost to the readers, because they must provide a dedicated receiver for every reception channel. On the other hand, tags should be able to distinguish between different frequencies and to select the sub-channels of interest. Only active tags add the previous functionality.

- Time Division Multiple Access (TDMA): The channel is divided into time slots that are assigned to the users. One of the most important problems of this technique is the users must be synchronized to send their information in the slot selected. This technique can be applied directly to RFID. For passive RFID systems, the tag's simplicity requires the reader controls the synchronization (centralized). For active RFID system, the synchronization can be centralized or the tags can control the synchronization themselves (distributed).
- Space Division Multiple Access (SDMA): This technique reuses certain resources, such as channel capacity in spatially separated areas. This technique can be applied to an RFID system as follows: In a scenario with two or more readers, the read range of each one is reduced but compensated by forming an array of antennas, providing then a large coverage area. The main drawback is the high implementation cost of the complicated array antennas system.
- Code Division Multiple Access (CDMA): It consists of using spread-spectrum modulation techniques based on a pseudo random code to spread the data over the entire spectrum. CDMA is the ideal procedure in many applications, e.g. navigation systems, GPS, etc. However, in RFID systems, this technique means more complex hardware in the tags and hence, higher cost.
- Carrier Sense Multiple Access (CSMA): This technique requires the tags to sense the channel traffic before sending their information. If there is no traffic, the tag starts to send. This mechanism can be only used with active tags because passive tags cannot monitor the channel.

Many of these solutions are not cost-effective due to the extra complexity of the tag. This is why solutions to the collision issue are usually sought at the MAC layer, tackling the burden of the algorithm at the reader equipment. Besides, new reactive procedures are being explored. Namely, those which extract useful information from colliding signals at Physical layer:

- The application of Radar-Cross-Section (RCS) has been proposed in the field of RFID by. The number of collided tags is detected by means of the analysis of the RCSs. Then, Minimum Distance Detector (MDD) mechanism is used to decode colliding signals. This technique may only be useful for collisions where only two tags are involved. Indeed, this mechanism has been only simulated using ISO 18000-6C as the underlying standard.
- Constellations analysis computes IQ constellations produced by additive simultaneous tag responses, and determines symbol decoding regions of transmissions.

This technique has been described for Low Frequency (LF) RFID systems in and for Ultra High Frequency (UHF).

Although these techniques are still immature, the authors point out their feasibility and compatibility with current standards and tags.

Anti-collision Protocols at MAC Layer

When a number of tags/readers are presented simultaneously in the coverage area an appropriate Medium Access Control (MAC) protocol is needed to handle/avoid collisions caused by simultaneous transmissions. Collisions in RFID occur in a number of ways:

- Case of a single reader-multiple tags collisions: Multiple tags are in the reading range of the same reader and respond simultaneously. The reader detects the electromagnetic wave but is unable to interpret the signal received.
- Case of multiple readers-single tag collisions: Only one tag is in the read range of multiple readers. The interferences occur when the signal from a neighboring reader collides with the tag transmission.
- Case of reader-reader collisions: Multiple readers configured to work within the same frequency band, interfere each other and thus a collision occurs.

Discussion of these three types of collisions would require a complete volume.

Tree-based Tag Anti-collision Protocols

Tree-based anti-collision protocols put the computational burden the reader. The reader attempts to recognize a set of tags in the coverage area in several interrogation cycles. Each interrogation cycle consists of a query packet, sent by the reader, and the response of tags in coverage. If a set has more than one tag, a collision occurs. When a collision occurs, the mechanism splits the set into two subsets using the tags identification numbers or a random number. The reader keeps on performing the splitting procedure until each set has one tag. Tree-based protocols are not efficient when the number of tags to recognize is large due to the lengthy identification delay.

Aloha Protocols

Aloha protocols are classified into four main groups. The first one is the Pure-Aloha protocol which is the simplest anti-collision scheme for passive tags with read-only memory. The second group is the Slotted Aloha protocol Slotted Aloha protocol is based on Pure-Aloha. A tag can transmit only at the beginning of a slot. Therefore, packets can collide completely or not collide at all. The mechanism is as follows: the reader sends a packet announcing the number of slots (K) that tags can compete to use. Each tag receives the data and generates a random number between [0, K-1]. The result is the slot where the tag must transmit their identification number.

Slotted-Aloha outperforms Pure-Aloha at the cost of requiring a reading system that manages slotted time synchronization. The third group, Frame-Slotted-Aloha (FSA), is a variation of Slotted-Aloha. In FSA, the time is divided into discrete time intervals but slots are confined in consecutive frames, also called cycles. Each frame has a length of a fixed number of slots. FSA has been implemented in many commercial products and has been standardized in ISO/IEC-18000-6C and in EPCglobal Class-1 Gen-2.

In FSA, when the number of tags is much larger than the number of slots, the identification delay increases considerably. On the other hand, if the number of tags is low and the number of slots is high, many empty slots can occur, which leads to increased identification time. In Dynamic FSA, the number of slots per frame is variable. Tags randomly choose a slot within the frame to send their information to the reader. When a frame finishes, an identification cycle concludes and the reader, following some rules, makes a decision about whether to increase/decrease/maintain the number of time-slots per frame in the next identification cycle.

The optimum throughput in a cycle of a DFSA protocol is achieved if the number of tags N equals the number of slots K in that cycle, and this throughput is given by $e^{-1} \approx 0.36$. Since the number of tags in range per cycle is commonly unknown, first the reader must estimate the number of tags that are going to compete per cycle, possibly through the statistical information collected on a cycle-by-cycle basis or any heuristic methods. Then, the reader adjusts the frame size to guarantee the maximum throughput and minimize the identification delay.

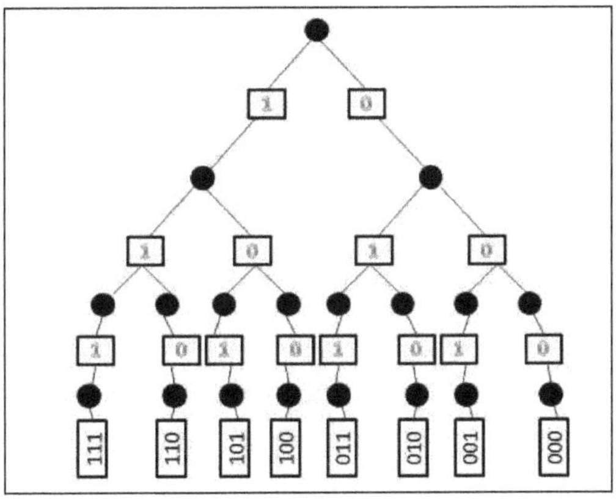

Tree-based protocol procedure.

EPCglobal Class-1 Gen-2

EPCglobal is an institution focused on the development of industry-driven standards for the Electronic Product Code (EPC) to support the use of Radio Frequency Identification (RFID). Regarding passive RFID, EPCglobal provides the EPCglobal Class-1 Gen-2 standard.

EPCglobal Class-1 Gen-2 is called "the worldwide standard for RFID systems" because it has been implemented to satisfy all the needs of the final customer, irrespective of the geographic location. For passive RFID systems, EPCglobal Class-1 Gen-2 is considered the de facto standard. It includes a set of specifications for the hardware of the passive tag and the hardware and software in the reader systems (which carry the true system complexity). It has been widely adopted by RFID systems manufacturers. Many commercial RFID systems have been implemented following this standard.

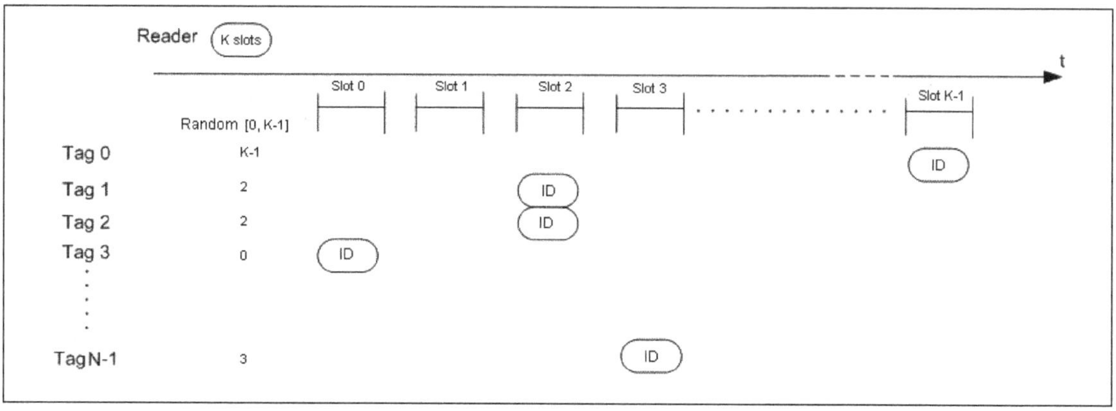

Frame Slotted Aloha procedure.

EPCglobal Class-1 Gen-2 works at UHF band (860MHz-930MHz). It proposes an anticollision mechanism based on a variation of FSA. Figure below illustrates EPCglobal Class-1 Gen-2 operation.

At a first stage the reader system is continuously monitoring the environment to detect the presence of tags by means of Broadcast packets. Tags in the coverage area are excited by the electromagnetic waves of the reader and send a reply immediately, producing a multiple collision. The reader detects the collision and starts the identification cycle. During each identification cycle, the time is structured as one frame, which is itself divided into slots, following a FSA scheme.

EPCglobal Class-1 Gen-2 shows two configuration alternatives:

- Fixed frame-length procedure: All identification cycles (frames) have the same value (number of slots). It is common to find commercial systems with this configuration.

- Variable frame length procedure (denoted as frame-by-frame adaptation). The number of slots per frame can be changed by the reader in each identification cycle. The reader decides if increase, decrease or maintain the number of slots per frame in function of some criteria.

Fixed Frame Length Procedure

An identification cycle starts when the reader transmits a Query packet, including a field of four bits with the value Q ∈ [0,...,15], stating that the length of the frame will be of 2Q slots. Tags in coverage receive this packet and generate a random number r in the interval [0, 2Q-1]. The r value represents the slot within the frame where the tag has randomly decided to send its identification number ID=r. Inside each frame, the beginning of a slot is governed by the reader by transmitting the QueryRep packet, excepting the slot 0, which is automatically initiated by the Query packet. The tags in coverage use an internal counter to track the number of transmitted QueryRep packets since the last Query packet, and then recognize the slot when they should transmit.

When the moment arrives, the tag transmits its identification number ID, which corresponds to the random value r calculated for contention, which is also equal to the slot number in the frame. After transmitting its ID, three actions can follow:

- If more than one tag has chosen the same slot, a collision occurs which is detected by the reader. Then, the reader reacts initiating a new slot with a QueryRep packet. The tags which transmitted their ID assume that a collision occurred, and must update their counter value to 2Q-1. That means that they will not compete again in this identification cycle.
- If the reader receives the ID correctly, and this coincides with the slot number within the frame, then it responds with an Ack packet. All tags in coverage receive the packet but only the identified tag answers with a Data packet, e.g. an EPC code. If the reader receives the Data packet, it answers sending a QueryRep packet, starting a new slot. The tag identified will finish its identification process.
- If the reader does not receive a correct Data packet within a given time, it considers the time-slot has expired, and sends a Nack packet. Again, all tags in coverage receive it, but only the tag in the identification reacts by updating its counter value to 2Q-1. Thus, this tag will not contend again in this identification cycle. After this, the reader will send a new Query or QueryRep packet to start a new frame or slot respectively.

Finally, when a cycle finishes, a Query packet is sent again by the reader to start a new identification cycle. Tags unidentified in the previous cycle will compete again, choosing a new random r value.

Variable Frame Length Procedure

The fixed frame length EPCglobal Class-1 Gen-2 standard provides a low degree of flexibility. If the Q value selected is high and the number of tags in coverage is low, many empty slots appear in the frame. On the contrary, if the Q value is low and the number of tags is high, many collisions arise. To mitigate this problem the standard proposes a variable frame length procedure that selects the Q value in each cycle by means of some arbitrary function.

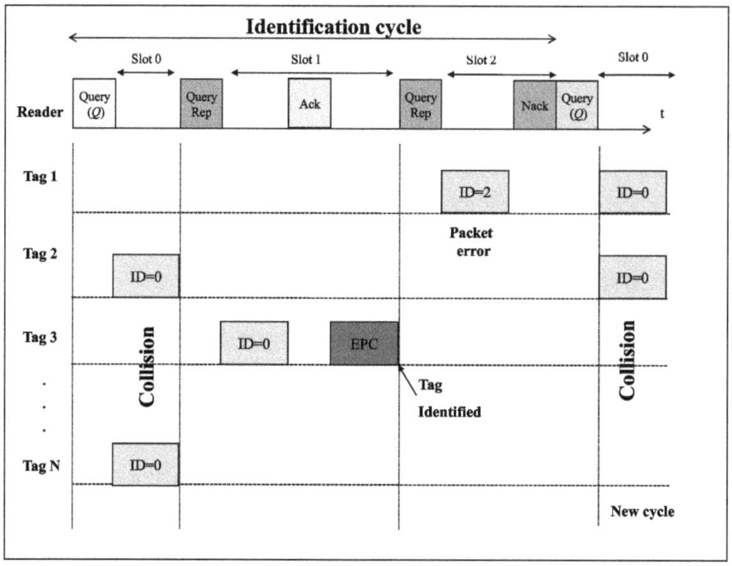

EPCglobal Class-1 Gen-2 identification procedure.

EPCglobal Class-1 Gen-2 in the Market

The current UHF RFID readers available in the market implement the worldwide standard EPCglobal Class-1 Gen-2. Some of them only permit to work with one of the two procedures explained before. Besides, some readers do not permit to configure the initial frame-length (the Q value) or only some certain values which can influence directly to the final system performance. Depending on the level of frame-length configuration, the readers can be classified as follows:

- Readers with fixed frame length, without user configuration (Symbol, on-line; ThingMagic, on-line; Mercury4, on-line; Caen, on-line; Awid, on-line; Samsys, on-line). Identification cycles are fixed and set up by the manufacturer. It is not possible to modify by the user (it is usually fixed to 16 slots). Therefore, these readers are not able to optimize the frame-length.
- Readers with fixed frame length with user configuration(Samsys, on-line; Intermec, online; Alien, on-line). Before starting the identification procedure the user can configure the frame length, choosing between several values, which depend on the manufacturer. Then, the identification cycle cannot be changed. If the user wants to establish a different value of frame-length, it is necessary to stop the identification procedure and restart with the new value of frame-length.
- Readers with variable frame length (Samsys, on-line; Intermec, on-line; Alien, on-line). The user only configures the frame- length for the first cycle. Then the frame-length is self-adjusted trying to adapt to the best value in each moment, following the standard proposal.

Identification Process in Static Scenarios

Static scenarios are characterized by a block of tags (modeling a physical pallet, box, etc.) that enter the checking area and never leave. Two related performance measures are commonly considered: The identification time, defined as the mean number of time units (slots, cycles, seconds, etc.) until all tags are identified, and the system throughput or efficiency, defined as the inverse of the mean identification time, i.e., the ratio of identified tags per time unit.

Markovian Analysis

The identification process in a static scenario is determined by the number of remaining unidentified tags. Thus, the identification process can be modeled as a homogeneous (Discrete Time Markov Chain) DTMC, X_c, where each state in the chain represents the number of unidentified tags, being c the cycle number. Thus, the state space of the Markov process is {N, N-1,..., 0}. Figure below shows DTMC state diagram from the initial state, X_0=N. The transitions between states represent the probability to identify a certain quantity of tags t or, in other words, the probability to have (N-t) tags still unidentified.

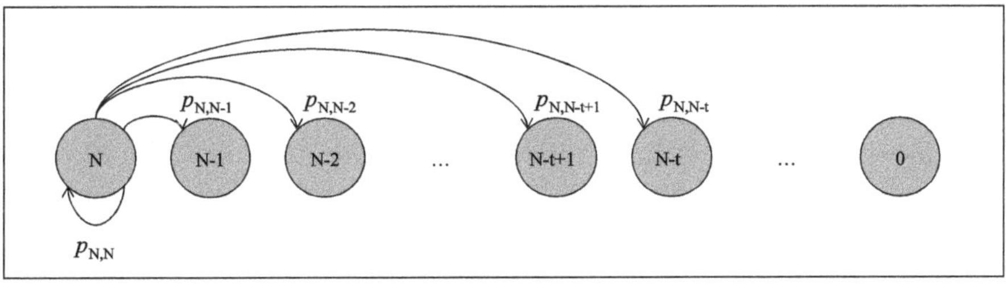

Partial Markov Chain.

The transition matrix P depends on the anti-collision protocol used and its parameters. For EPCglobal Class-1 Gen-2, the parameter K denotes the number of slots per frame (frame length). To compute the matrix P, let us define the random variable μ_t, which indicates the number of slots being filled with exactly t tags. Its mass probability function is:

$$\Pr_{K,N}(\mu_t = m) = \frac{\binom{K}{m}\prod_{i=0}^{m-1}\binom{N-it}{t} G(K-m, N-mt, t)}{K^N}$$

Where $m=0,\ldots,K$ and:

$$G(M,l,v) = M^l = \sum_{i=1}^{\left\lfloor \frac{l}{v} \right\rfloor}\left\{(-1)^i \prod_{j=0}^{i-1}\left\{\binom{l-jv}{v}(M-j)\right\}(M-i)^{l-iv}\frac{1}{i!}\right\}$$

Since the tags identified in a cycle will not compete again in the following ones, then the transition matrix P is:

$$p_{i,j} = \begin{cases} Pr_{K,i}(\mu_1 = i = -j) & ,i-K \le j < i \\ 1 - \sum_{y=i-1}^{i-K} P_{i,y} & ,i = j \\ 0 & ,\text{otherwise} \end{cases} \quad \text{for } i = 1,\ldots,N.$$

The chain has a single absorbing state, X_c=0. The mean number of steps until the absorbing state is the mean number of identification cycles $(\bar{c})$. It can be computed by means of the fundamental matrix, D, of the absorbing chain:

$$D = (I - F)^{-1}$$

As usual, I denotes the identity matrix, and F denotes the submatrix of P without absorbing states. Then,

$$(\bar{c}) = \sum_{j \in B} D_{Z,j}$$

Where *B* is the set of transitory states, and *Z* is the absorbing state.

In addition, using the physical and FSA standard parameters is possible to transform the identification time to seconds as follows: T_{id} is the duration of a slot with a valid data transmission (EPC code). T_v and T_c is the duration of an empty and collision slot, respectively. Then, the identification time in seconds is approximated by:

$$\bar{T}_{total} \approx \bar{c} \cdot \left[\bar{k}_v \cdot T_v + \bar{k}_c \cdot T_c + \bar{k}_{id} \cdot T_{id} \right]$$

$\bar{k}_V, \bar{k}_c$ *and* $\bar{k}_{id}$ denote the average number of empty, collision and successful slots, respectively. These variables depend on the particular FSA algorithm and its configuration, and on the population size. For instance, setting M=4, T_{id}=2.505 ms and T_v=T_c=0.575 ms. Since an empty slot and a collision slot have the same duration, the previous equation can be simplified:

$$\bar{T}_{total} \approx \bar{c} \cdot \left[\left(\bar{k}_v + \bar{k}_c \right) T_c + \bar{k}_{id} \cdot T_{id} \right]$$

Since,

$$\bar{k}_v + \bar{k}_c \approx K \cdot \bar{c} - \bar{k}_{id}$$

Then,

$$\bar{T}_{total} \approx \bar{c} \cdot \left[\left(K \cdot \bar{c} - \bar{k}_{id} \right) T_c + \bar{k}_{id} \cdot T_{id} \right]$$

Different populations of tags and Q values have been considered and the identification time has been measured. Figure shows the mean number of slots required to identify each tag population.

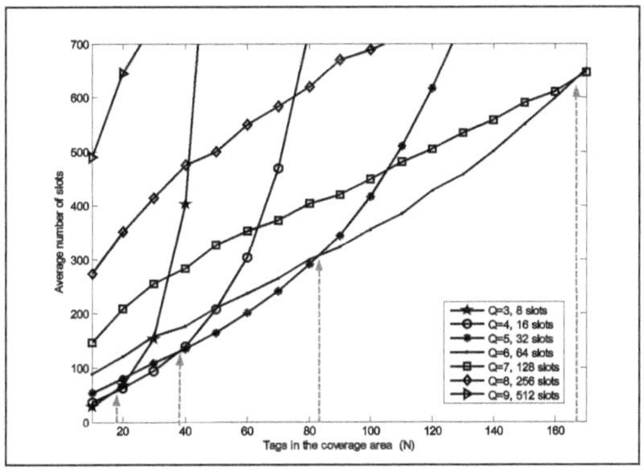

Mean identification time (in number of slots) vs. N, for different Q values.

System Throughput

The throughput (th) can be computed from the previous Markov analysis, just as the inverse of the identification time. Another way is described. Let us remark that, obviously, the result of both methods is equal, and the second one is provided for completeness. Given N tags, and K slots, the probability that t tags respond in the same time-slot is binomially distributed:

$$\Pr(t)=\binom{N}{t}\left(\frac{1}{K}\right)^{t}\left(1-\frac{1}{K}\right)^{N-t} \quad \text{for } t=0,\ldots,N$$

Then, Pr(t=0) is the probability of an empty slot, Pr(t=1) the probability of a successful slot, and Pr(t ≥2) the probability of collision:

$$\Pr(t=0)=\left(1-\frac{1}{K}\right)^{N}$$

$$\Pr(t=1)=\frac{N}{K}\left(1-\frac{1}{K}\right)^{N-1}$$

$$\Pr(t\geq 2)=1-\Pr(t=0)-\Pr(t=1)=1-\left(1-\frac{1}{K}\right)^{N}\left(1-\frac{N}{K-1}\right)$$

Since every identification cycles is composed by K slots, the throughput per slot is computed as follows:

$$th=K\cdot\Pr(t=1)=N\left(1-\frac{1}{K}\right)^{N-1}$$

Identification Process in Dynamic Scenarios

Many real RFID applications (e.g. a conveyor belt installation) work in dynamic scenarios. For this type of systems, the performance analysis must be linked with the Tag Loss Ratio. This parameter measures the rate of unidentified tags in an identification process and, depending on the final application, even a low TLR (e.g. TLR=10^{-3}) may be disastrous and cause thousands of items lost per day. The TLR is computed for a RFID scenario similar to the one. There is an incoming flow of tags entering the coverage area of a reader (RFID cell), moving at the same speed (e.g., modeling a conveyor belt). Therefore, all tags stay in the coverage area of the reader during the same time. Every tag unidentified during that time is considered lost. Once acknowledged, a tag withdraws from the identification process. Thereafter, the following notation and conventions are used: a row vector is denoted as $\vec{V}$, the i-th component of a vector is denoted $\left(\vec{V}\right)_{i,}$ and $\sigma\left(\vec{V}\right)$ denotes the sum of the values of the components of a vector $\vec{V}$.

For the sake of simplicity, let us assume tags remain C complete cycles in the reading area. Then, once a tag has entered the coverage area, it should be identified in the following C identification cycles. Otherwise (if it reaches the cycle C+1), tag is lost. A truncated Poisson distribution, with parameter λ, has been selected as the arrival process in the system:

$$a(t)=\frac{\lambda^t}{t!\sum_{i=0}^{H}\frac{\lambda^t}{i!}}$$

For t=,...,H, being H the maximum number of tags entering per cycle. The former assumptions allow to express the dynamics of the system as a discrete model, evolving cycle by cycle, such that,

- Each tag is in a given reading cycle in the set [1,...,C].
- After a cycle, identified tags withdraw from the identification process.
- After a cycle, each tag unidentified and previously in the i-th cycle moves to the (i+1)-th cycle.
- If a tag enters cycle C+1, it is considered out of the range of the reader, and, therefore, lost.
- At the beginning of each cycle, up to H new tags are assigned to cycle 1, following a truncated Poisson distribution.

For any arbitrary cycle, the evolution of the system to the next cycle only depends on the current state. Thus, a DTMC can be used to study the behavior of the RFID system.

The system can be modeled by a homogeneous discrete Markov process X_c, whose state space is described by a vector $\vec{E} = \{e_{1,\dots,} e_{C+1}\}$, where each $e_j \in [0,\dots,H]$, representing the number of unidentified tags in the j-th cycle. The following figures illustrate the model. They describe the state of the system for two consecutive cycles, showing tags entering and leaving the system, in both identification and no identification scenarios. Therefore, e_j is the number of tags which are going to start their j-th identification cycle in coverage. e_1 component also represents the number of tag arrivals during the previous identification cycle (which do not contend since they have not received a Query packet yet). Finally, component e_{C+1} indicates the number of tags lost at the end of the identification cycle, since tags leave coverage area after C+1 cycles. In addition, let us define the mapping Ψ as a correspondence between the state vector and an enumeration of the possible number of states:

$$\psi : [0,\dots,H] \times^{(C+1)} \times [0,\dots,H] \rightarrow \left[1,\dots,(H+1)^{C+1}\right]$$

$$\vec{E} = \{e_{1,\dots,} e_2,\dots,e_{C+1}\} \rightarrow \psi : (\vec{E}) = 1 + \sum_{j=1}^{C+1} e_j H^{j-1}$$

This allows defining i-th state in our model as the state whose associated vector is given by Ψ^{-1}. Let us denote $\vec{E}_i$ as the vector associated to i-th state, i.e., $\vec{E}_i = \psi^{-1}$ (i). Finally, let e_{ij} denote to the j-th component of the $\vec{E}_i$ state vector.

The goal is to describe the transition probability matrix P for the model, from every state i to another state j. The stationary state probabilities is computed as $\vec{\pi} = \vec{\pi}$ P. Let us denote λ_j as the average incoming unidentified tags to cycle j, which can be computed as:

$$\lambda_j = \sum_{j=1}^{(H+1)^{C+1}} e_{ij} \vec{\pi}_i$$

Obviously, λ_1 is the average incoming traffic in the system and λ_{C+1} is the average outgoing traffic of unidentified tags. Then, TLR can be calculated as:

$$TLR = \frac{\lambda_{C+1}}{\lambda_1} = \frac{\sum_{i=1}^{(H+1)^{C+1}} e_{i(C+1)} \cdot \vec{\pi}_i}{\lambda_1}$$

To build the transition probability matrix P let us define the auxiliary vectors $\vec{L}_i$ and $\vec{U}_i$ as:

$$\vec{L}_i = \{e_{i1},\dots,e_{iC}\}$$

$$\vec{U}_i = \{e_{i2},\dots,e_{i(C+1)}\}$$

That is, the $\vec{E}_i$ state vector without either the last or the first component. Let us define the outcome vector as:

$$\vec{O}^{ij} = \left(\vec{L}_i - \vec{U}_j\right) = \left\{o_1^{ij} - o_C^{ij}\right\}$$

Figures graphically show this computation. To construct the transition matrix let us define the function id(i,j) that operates on an outcome vector $\vec{O}^{ij}$ providing the number of identified tags in a transition from a state i to a state j:

$$id(i,j) = \vec{O}^{ij} \cdot \vec{1}'$$

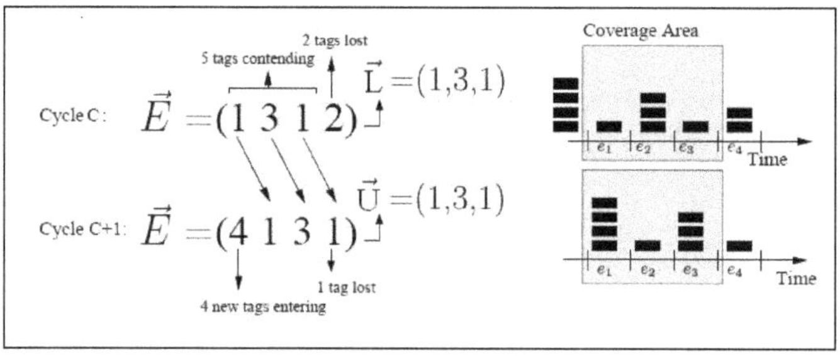

Representation of the transition state. Case 1: No identification.

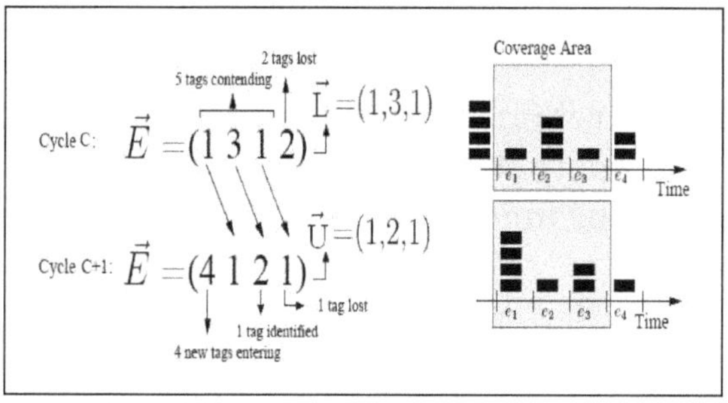

Representation of the transition state. Case 2: Identification.

For $\vec{E}_i$ and $\vec{E}_j$, if $e_{ik} < e_{j(k+1)}$ for some k=1,...,C, such transition is impossible (new tags cannot appear in stages other than stage 1). These impossible transitions will result in id(i,j) providing a negative value. The random variable s(K,N) indicates the number of contention slots being filled with a single tag. The mass probability function of s(K,N). Henceforth, let us denote Pr{s(K,N)=k} as s_k=(N,K).

Using FSA, up to K tags may be identified in a single identification cycle. Therefore, possible cases range from id(i,j)=0 to id(i,j)=K. The probability of id(i,j) successful identifications is uniformly distributed among the contenders, whose distribution depends on the particular state, and hence the transition probability. From equations $\Pr_{K,N}\left(\mu_t = m\right) = \frac{\binom{K}{m}\prod_{i=0}^{m-1}\binom{N-it}{t}G(K-m, N-mt, t)}{K^N}$ and

$$G(M,l,v) = M^l = \sum_{i=1}^{\left\lfloor \frac{l}{v} \right\rfloor} \left\{ (-1)^i \prod_{j=0}^{i-1} \left\{ \binom{l-jv}{v} (M-j) \right\} (M-i)^{l-iv} \frac{1}{i!} \right\}$$ and the previous definitions, the transition matrix P can be computed as follows:

$$P_{i,j} = \begin{cases} a(e_{j1}) s_0 (K,N) & ,id(i,j)=0 \\ a(e_{j1}) \dfrac{\prod_{k=1}^{C} \binom{e_{ik}}{o_k^{ij}}}{\binom{N}{id(i,j)}} S_{id(i,j)}(K,N) & ,id(i,j) \in [1,K] \\ 0 & ,\text{otherwise} \end{cases}$$

Experimental Evaluation: A Postal Mail Control System

From a practical point of view, TLR evaluation may become critical in some realistic scenarios. Two configurations for the mail sojourn times of 2 and 3 identification cycles have been considered, for a frame length of K=8 slots. The slot time is assumed to be 4 ms based on parameters. Therefore, the time sojourn is around 64 ms for C=2 and 100 ms for C=3. λ range spans from 1 to 7. Results are provided in figures. As expected, for a fixed C, TLR increases as the maximum number of arrivals H increases. In addition, for the parameters analyzed, keeping fixed H decrements TLR if C grows, because there are more opportunities for identification. For example, the maximum number of tags for H = 6 and C=2 is 12 tags, whereas for H = 6 and C=3 there might be up to 18 tags.

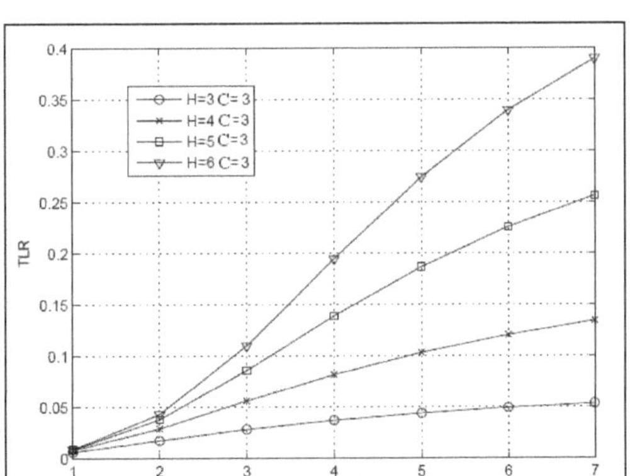

TLR results for FSA with 8 slots and Poisson arrivals. C=4, and H=3 to H=6.

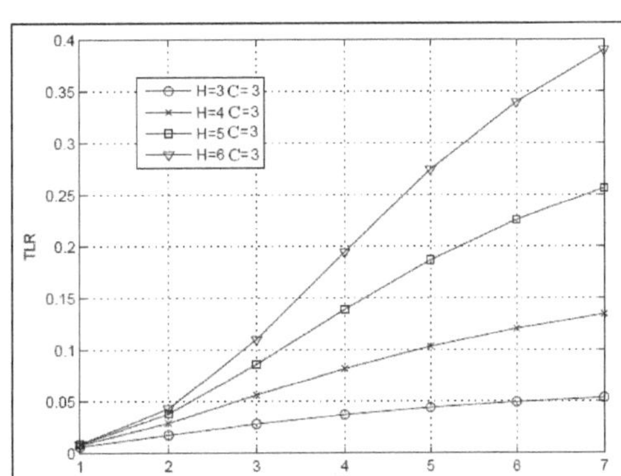

TLR results for FSA with 8 slots and Poisson arrivals. C=5, and H=3 to H=6.

The main issue of the previous analysis is that it becomes computationally unfeasible for moderate values of H and C. In this case, simulation is mandatory. Figure below shows simulations performed for λ=[10,...,60] and H = [3; 6]. In this case, envelopes sojourn time is close to 800ms. We can observe that, if we set H=3, TLR reaches 10^{-4} and does not vary, independently of the λ value. On the other hand, with H=6, the TLR reaches up to 10^{-3}. It means that, one out of a thousand envelopes will be lost, showing the impact in the final system.

The evaluation of TLR for different protocol parameters, such as the number of slots, the arrival process, the time in coverage (conveyor belt velocity), etc.

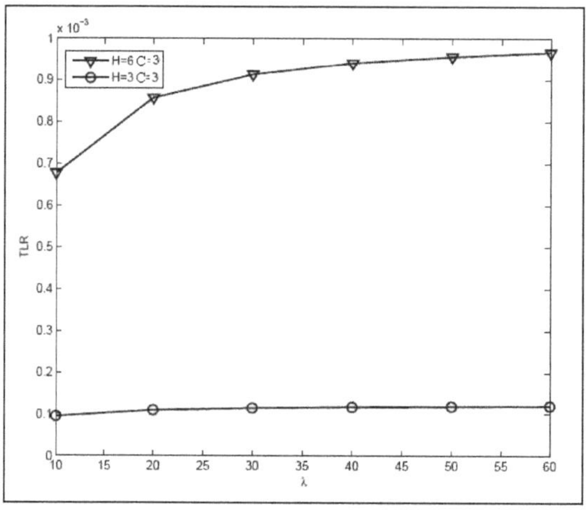

TLR results for FSA with 64 slots and Poisson arrivals, C=4 and H=3, H=6.

MAC PROTOCOLS FOR RFID SYSTEMS

An RFID system is formed by a set of n tags, that represent clients in the identification process, and a tag reader, that is the identification server. Each tag is associated with an identifier (ID) that must be transmitted to the reader upon receipt of a suitable identification query. Tags can be active, i.e. provided with an independent power supply (like a battery), or passive. In the latter case, tags depend on energy provided by the tag reader through its queries. For this reason, active tags can have some data processing ability, while passive tags are limited to doing only elementary operations and replying to the reader's queries.

An important issue in Radio Frequency Identification systems concerns the reading of colocated tags that must be managed through suitable Medium Access Control (MAC) protocols, specifically designed for low power devices. Efficiency of MAC protocols for RFID systems directly influences the time needed by the tag reader for completely identifying a set of co-located tags, so it has great impact on the whole system performance.

We denote by $T_{tot}(n)$ the time needed by the tag reader for completely identifying a set of n tags. Under ideal conditions, it should be $T_{tot}(n) = nT_{pkt}$, where T_{pkt} is the time needed to transmit one identification packet. However, in order to ensure collision-free transmission to each tag, a MAC protocol is needed that, inevitably, produces a time waste with respect to the ideal condition. The time actually needed for identification depends on several factors, like the technology adopted, the working frequency, etc. So, RFID systems can exhibit very different reading rates, typically ranging between 100 Tag/s and 1000 Tag/s.

The most common MAC protocols used in RFID systems can be grouped into two classes: deterministic protocols and stochastic protocols. Deterministic protocols basically coincide with tree traversal algorithms: all RFID tags form a binary tree on the basis of their identifiers, and the reader explores the tree in a systematic way, by repeating queries based on bit masks. Randomness is only in the tree structure (due to the choice of the co-located tag set), while the algorithm execution is pre-determined.

Stochastic MAC protocols are instead based on the framed slotted Aloha (FSA) algorithm, that requires each tag to make a constrained pseudo-random choice of an integer number in order to reduce the probability of collisions. Both deterministic and stochastic MAC protocols for RFID systems have some advantages and disadvantages. Binary tree (BT) protocols have the advantage of very low tag complexity, since there is no need of implementing pseudo-random number generators within tags (except for singulation, when tags with the same identifier can be co-located). Furthermore, binary tree does not require the estimation of the number of co-located tags, that yields an additional effort before reading.

On the other hand, though needing a quite accurate estimation of the tags number, the FSA protocol can be more efficient than the binary tree protocol for large sets of co-located tags. Moreover, FSA does not need any particular bit coding of the tags identifiers and is intrinsically able to resolve ambiguity in the case of multiple tags with the same identifier.

Another important issue in RFID systems concerns power consumption that influences the tags' lifetime. In fact, the same (possibly short) identification time can be achieved at the cost of many collisions or with a small number of collisions. In the latter case, power consumption is obviously minimized, with the effect of a prolonged tags' lifetime. In general terms, stochastic protocols are recognized to be more power efficient with respect to deterministic ones, due to the reduced number of collisions they produce.

Both deterministic and stochastic MAC protocols have been included in standards and specifications for RFID systems. For example, EPCglobal Class 0 and Class 1 Generation 1 specifications adopt two different binary tree algorithms for the reading of co-located tags, and the same occurs in the first versions of the ISO 18000-6 standard for RFID systems. The more recent EPCglobal Class 1 Generation 2 specification introduced a new inventory technique based on the FSA protocol, that has also been included in the ISO 18000-6 Type C standard.

Several alternative solutions have been proposed in the literature for the implementation of anti-collision algorithms targeted to RFID applications. For many practical and commercial applications, however, a fundamental requirement is that RFID tags must be very simple and inexpensive devices, often designed for single use. So, it is of main importance to develop low complexity anti-collision protocols able to solve the issues related to the shared medium while considering power and cost constraints.

The Binary Tree Protocol

The binary tree protocol is one of the most simple arbitration protocols, and it is based on the random splitting of the whole group of clients into two subgroups each time a collision occurs. This way, clients are progressively separated into smaller groups, until no more than a single client remains in each group. This is equivalent to put the clients on the nodes of a tree having maximum nodal degree 2. In the binary tree protocol implementations used for medium access control in RFID systems, the binary splitting of tags into subgroups is based on the value of their IDs, that have fixed length and are supposed to be unique.

The tag reader is responsible for the management of the tree traversal procedure. The algorithm starts when the tag reader announces the tree traversal and transmits a binary value. All tags having that value as the first bit of their ID reply by transmitting back the same value, while the other tags exit from the traversal and wait for another query (with different initial value).

The reader receives the transmitted values for the next bit, and it may or not detect a collision. In fact, it may happen that some tags transmit the same value, but the reader is not able to distinguish if transmission was from a single tag or more than one. In both cases, when a reply is received, the reader goes on with the binary splitting by choosing and transmitting a binary value for the next bit, until completion of the tag ID. Actually, according to the ISO 18000-6 standard, sending a null value corresponds to no transmission that helps to reduce the power consumption.

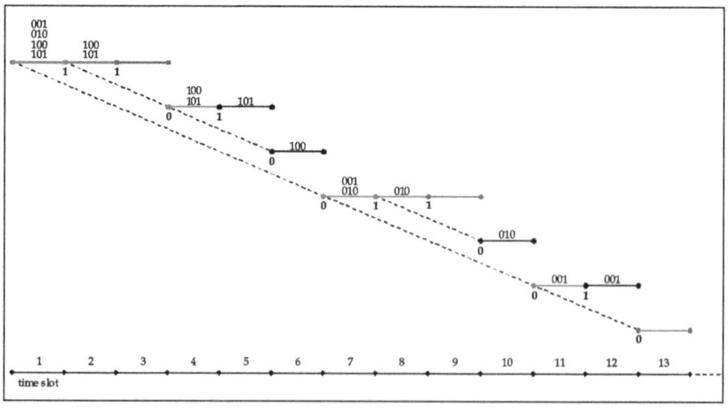

Example of binary tree protocol with n = 4 tags.

An example of binary tree protocol used for reading co-located RFID tags. We suppose there are n = 4 tags with the following identifiers: 001, 010, 100 and 101. The time axis is divided into time slots; each time slot begins when the tag reader makes a new query for the value of a bit. The first time slot begins with the announcement of a multiple read query by the tag reader. All tags reply when such command is issued. The identifiers of the tags responding to each query are reported in the time slot that follows the query.

After the first time slot, each tag replies to the query if and only if the next bit of its ID coincides with the queried value; otherwise, it goes to sleep mode. When more than one tag reply in the same time slot, a collision occurs and the corresponding time slot is marked in red. When the bit queried is not the last one, the tag reader is not able to tell a single reply from a collision, and must continue with the traversal. So, single replies at all bits except the last are equivalent to collisions (this occurs for the tags with IDs 010 and 001 in the figure). When no tag replies to a query, we have an idle time slot, marked in blue in the figure. Successful identification is instead accomplished when the last bit is queried and only one tag replies. The corresponding time slots are marked in black in the figure and labelled with the corresponding tag identifier.

An important feature of the binary tree protocol exemplified in figure is that queries that are interrupted are then restarted exactly from the same point. In other terms, the tag reader must be able to store the status of each branch of the tree. This way, each edge of the tree is travelled only once, and queries are never restarted from the beginning. This version of the algorithm is with memory, and it is opposite to memoryless versions, whose efficiency is reduced due to repeated queries within the traversal.

Due to its simplicity, the binary tree protocol is suitable for being modelled analytically, and theoretical arguments can be used to predict the value of its most important parameters, that are: the number of tree levels required for a random contender to have success, the total number of tree levels and the number of contention frames required to complete the algorithm. Other relevant parameters, as throughput and delay, can also be estimated through analytical modelling of the protocol.

The total number of tree levels obviously depends on the number of clients n, due to the assumption that, at the lowest tree level, each node must be associated, at most, to one client. The mapping between clients and tree nodes is stochastic, so the allocation of the tree nodes is not optimal. The total number of tree levels (D_n) can be lower bounded by considering the optimal distribution of clients on tree nodes. This occurs when even size groups of tags are split into equal subgroups, while odd size groups are split into subgroups having sizes that differ by one. In this case, it is simple to observe that:

$$D_n \geq \lceil \log_2(n) \rceil + 1,$$

where function gives the smallest integer greater than or equal to its argument. In practice, due to the statistic nature of collisions, the number of levels is usually higher. In it is proved that, for large n, the average number of tree levels is,

$$D_n \cong 2\log_2(n).$$

Another important parameter to evaluate the time needed by the binary tree protocol to complete the identification and to be compared with other arbitration protocols is the total number of time slots as a function of n. In order to estimate it for finite values of n (not necessarily large), we can resort to some simple theoretical arguments.

First, we consider that each collision generates two edges, corresponding to the two possible choices for the bit under analysis. So, the total number of time slots required by the binary tree (I_{tot}) to completely identifying a set of n tags is simply twice the number of collisions (C_{bin}), augmented by one (to consider the first time slot):

$$I_{tot}(n) = 2C_{bin}(n) + 1$$

If we focus on the i-th level of the tree, the number of time slots is $m = 2^i$, and each tag must reply to a query in one of them. The probability that a tag does not reply in a time slot is:

$$p(m) = 1 - \frac{1}{m}$$

and the probability that the time slot is idle (that is, no tag replies to a query in that time slot) is $p(m)^n$. So, the average number of idle time slots at level i is:

$$Q(n,m) = m \cdot p(m)^n = m \cdot \left(1 - \frac{1}{m}\right)^n.$$

We have reported in figure the value of Q(n, m) expressed by the above equation, as a function of the tree level (i), for different values of the number of tags (n). As expected, the average number of idle time slots quickly converges to the total number of time slots, that is, 2^i.

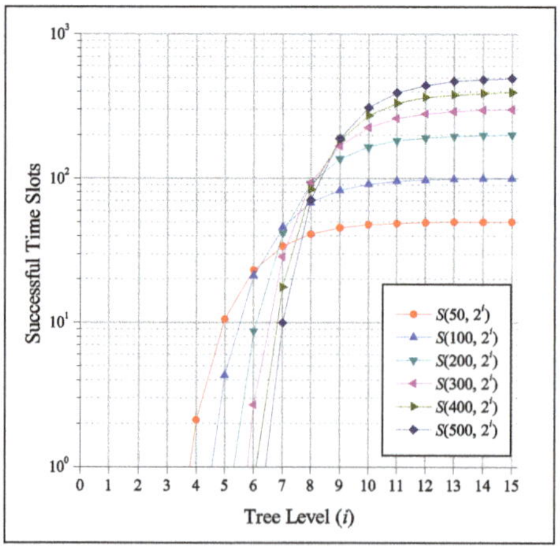

Average number of idle time slots as a function of the tree level.

With similar arguments, we can consider that a time slot at level i is successful if it is used by a single tag to reply to a query. So, the average number of successful time slots at the tree level i can be estimated as follows:

$$S(n,m) = m \cdot n \cdot p(m)^{n-1}\left[1-p(m)\right].$$

Figure reports the average number of successful time slots as a function of the tree level, for the same choices of the number of tags, calculated by means of the above equation. In this case, it is immediate to observe that the number of successful time slots converges to the total number of tags n.

Starting from the expressions $Q(n,m) = m \cdot p(m)^n = m \cdot \left(1-\frac{1}{m}\right)^n$ and $S(n,m) = m \cdot n \cdot p(m)^{n-1}\left[1-p(m)\right]$. for Q(n, m) and S(n, m), respectively, the average number of collisions at the tree level i can be estimated as the number of non-idle and nonsuccessful time slots, i.e.:

$$C_{bin}(n,m) = m - Q(n,m) - S(n,m).$$

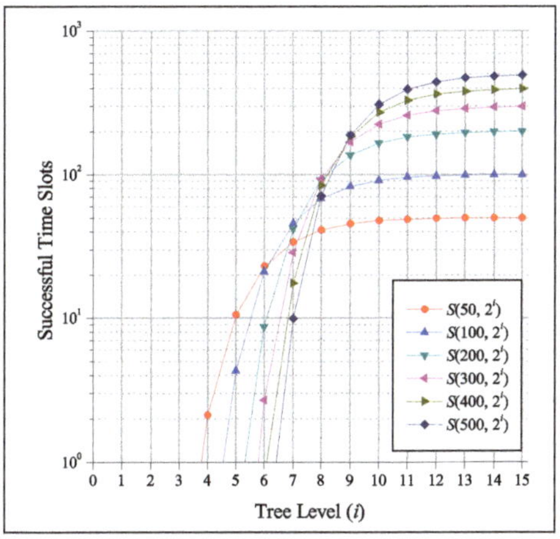

Average number of successful time slots as a function of the tree level.

Based on these considerations, it follows that the average number of time slots with collisions rapidly goes to zero (a negative number of collisions obviously has no sense). For better evidence, the average number of collisions, expressed by $C_{bin}(n,m) = m - Q(n,m) - S(n,m)$, for the same choices of n, as a function of the tree level. We observe that, for the initial tree levels, the number of time slots with collisions coincides with the total number of time slots (2^i). Then, the number of collisions becomes smaller than 2^i and, after reaching a maximum value, begins to decrease monotonically.

The total number of collisions in the binary tree protocol can be found by summing the average number of collisions at each tree level, that is:

$$C_{bin}(n) = \sum_{i=0}^{\infty} m\left[1+(n-1)p(m)^n - n \cdot p(m)^{n-1}\right].$$

Expression above allows to determine analytically the total number of collisions and, through equation $I_{tot}(n) = 2C_{bin}(n) + 1$, we can then estimate the total number of time slots needed by the binary tree algorithm to completely identifying the set of n tags. As we will see in the following, such analytical estimation gives results that are very close to those of numerical simulations.

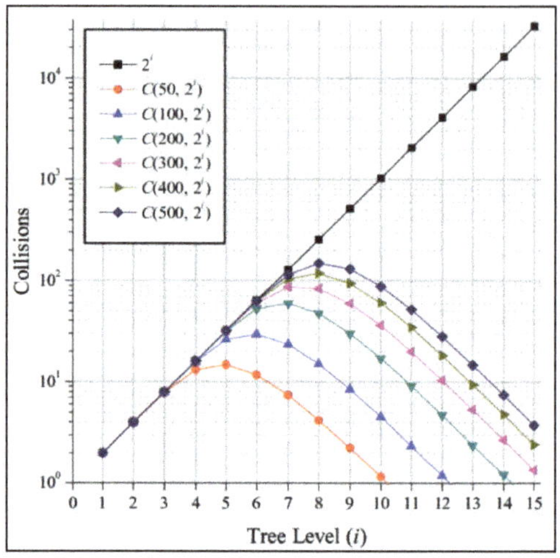

Average number of time slots with collisions as a function of the tree level.

The Framed Slotted Aloha Protocol

The binary tree protocol for medium access control in RFID systems exploits binary splitting of the tags into subgroups on the basis of their identifiers, that are fixed and known a priori. For this reason, the binary tree and other similar deterministic protocols are opposed to stochastic protocols, mostly based on the framed slotted Aloha algorithm.

In slotted Aloha protocols, the time axis is divided into time slots. Each tag synchronizes its transmission with the beginning of a time slot, in such a way that concurrent transmissions collide completely. This is the main difference between slotted Aloha protocols and the pure Aloha one, in which instead the time axis is not discretized, so partial collisions can occur as well. In the framed version of the slotted Aloha protocol, time slots are grouped into groups of L, and each group coincides with a frame. Each tag can transmit only once in each frame, and frames are repeated until the end of the identification procedure.

At the beginning of a frame, each tag randomly selects a time slot within the frame for transmitting its ID. The tag then transmits at the chosen time slot; if a collision occurs, colliding tags wait the end of the current frame and repeat the procedure in the following frame. The advantage of the introduction of frames is due to the limitation in the transmission rate imposed by the fact that each tag only transmits once in a frame. This allows to reduce the number of collisions in the initial phase of the protocol, when all tags try to communicate. This can result in a significant performance improvement with respect to slotted Aloha, on condition that the frame length is properly chosen.

In standard FSA, the frame length must be fixed a priori and is kept constant until completion of the algorithm. When the number of tags significantly exceeds the frame size, efficiency of the FSA protocol with fixed frames decreases. On the other hand, the algorithm efficiency could be kept high by adjusting the frame length on the basis of the number of active tags. In this case, however, such number must be estimated, that instead is not necessary in classic FSA.

A first solution to the problem of estimating the number of tags in the FSA protocol is to adopt dynamic versions of the FSA. These approaches exploit the dependence of the probability of collision on the frame size and the number of tags. Such dependence can be expressed in analytical terms through theoretical arguments similar to those used in the previous section for the analysis of the binary tree protocol. Let P_{idle}, P_{succ} and P_{coll} represent the probability that a time slot is idle, used for a successful transmission or occupied by a collision, respectively. Similarly to equation $Q(n,m) = m \cdot p(m)^n = m \cdot \left(1 - \frac{1}{m}\right)^n$ and $S(n,m) = m \cdot n \cdot p(m)^{n-1}\left[1 - p(m)\right]$ we can express P_{idle} and P_{succ} as follows:

$$\begin{cases} P_{idle} = \left(1 - \frac{1}{L}\right)^n \\ P_{succ} = n\frac{1}{L}\left(1 - \frac{1}{L}\right)^{n-1} \end{cases}$$

Starting from the above equation, P_{coll} can be calculated as:

$$P_{coll} = 1 - P_{idle} - P_{succ},$$

so n can be estimated from the knowledge of L and the estimate of P_{coll}. However, it has been observed that the estimate of n so obtained can be inaccurate, since, for high values of the collision probability, small errors in the estimation of P_{coll} may produce significant deviations of the estimated n from its actual value.

Another important result that can be derived from equation $\begin{cases} P_{idle} = \left(1 - \frac{1}{L}\right)^n \\ P_{succ} = n\frac{1}{L}\left(1 - \frac{1}{L}\right)^{n-1} \end{cases}$ is that the optimal frame size in the FSA algorithm exactly coincides with the number of tags n. So, FSA becomes less and less efficient when the gap between L and n increases.

For this reason, in Dynamic FSA (DFSA), the probability of collision is used to obtain an estimate of the number of tags that try to access the shared medium. This is repeated at each frame, in such a way that the frame length is dynamically adjusted on the basis of the actual number of contending tags. As anticipated, the estimated number of tags can be inaccurate. We will denote by α the ratio between the estimated number of tags and its exact value (thus, α = 1 represents the ideal behaviour).

FSA with Robust Estimation and Binary Selection

An efficient approach for estimating the number of contending tags in the FSA protocol has been proposed, where the variant denoted as "FSA with Robust Estimation and Binary Selection", or EB-FSA, has been introduced.

The EB-FSA protocol begins with an estimation phase that has the purpose of estimating the number of tags n and to adjust the frame size L consequently. The estimation phase proposed is robust, in the sense that it solves the issues due to high sensitivity of the estimated n to estimation errors on Pcoll, in equation $P_{coll} = 1 - P_{idle} - P_{succ,}$ for high values of the collision probability.

Estimation in EB-FSA starts by fixing an estimation frame size L_{est} and a target P_{coll} threshold ($P_{coll\text{-}th}$), that corresponds to a threshold number of tags (n_{th}). The tag reader estimates the value of n only when P_{coll} becomes smaller than $P_{coll\text{-}th}$, in such a way to reduce inaccuracy on the estimate of n (nest). For this purpose, at each iteration of the estimation phase, the tag reader reduces the number of tags polled by a factor f_d, using a bit mask in the query frame.

Estimation of the tags number is made only when P_{coll} becomes smaller than $P_{coll\text{-}th}$, i.e., after a number of estimation frames equal to i*(n), that coincides with the smallest i such that $n/f_d^{i-1} < n_{th}$. In formula:

$$i^*(n) = \underset{\substack{i \in N \\ \frac{n}{f_d^{i-1}} < n_{th}}}{\arg\max} \left(\frac{n}{f_d^{i-1}} \right)$$

Thus, in the EB-FSA protocol, the initial estimation phase requires $I_{est} = i^*(n)L_{est}$ time slots. After such initial phase, tags randomly choose an integer between 1 and nest for transmission of their ID, and a frame with length n_{est} is transmitted. Differently from FSA, when a collision occurs, colliding tags do not wait for the next frame to retransmit their ID. On the contrary, a binary selection mechanism is implemented, that works as follows:

- Non colliding tags increment their counters by 1.
- Colliding tags randomly choose a binary value.
- Colliding tags that chose 0 try retransmission at the next time slot.
- Colliding tags that chose 1 try retransmission after one time slot.
- If another collision occurs, the procedure is repeated within the new set of colliding tags.

The binary selection mechanism avoids the need for subsequent frames, since each collision is necessarily solved through the splitting procedure. The protocol succeeds when all the tags have been identified, that is, after I_{iden} time slots. So, the total number of time slots needed by the EB-FSA protocol for completing its task is:

$$I_{tot} = I_{est} + I_{iden} = i^*(n)L_{est} + I_{iden}.$$

As it will be evident through numerical simulations, the EB-FSA approach can achieve a significant performance improvement with respect to FSA and DFSA.

Protocols Comparison

In order to assess and compare the considered protocols for medium access control in RFID systems, we report in this section the results of numerical simulations that model different scenarios.

We are interested in estimating the time needed by the considered protocols to complete the identification phase, in order to compare their efficiency in arbitrating the channel use in groups of tags with different size. The actual identification speed depends on technology issues, so we refer to the number of time slots instead of real time. We consider, as a starting point, the classic implementation of the framed slotted Aloha protocol, with fixed frame size. We consider two common values of frame size, that are L = 128 and L = 256, and estimate by simulation the total number of time slots needed to complete the identification procedure (I_{tot}). The results obtained are averaged over a number of simulations sufficiently high to ensure a satisfactory level of statistic confidence.

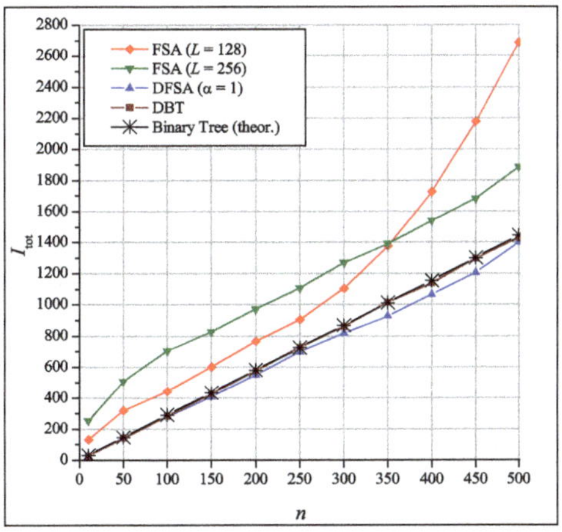

Comparison of binary tree and framed slotted Aloha protocols.

The values of Itot, so obtained as a function of the total number of tags (n). As we observe from the figure, the classic FSA protocol becomes less and less efficient for an increasing number of tags. The curve corresponding to L = 128 exhibits a parabolic behaviour starting from n on the order of 300. When L is increased up to 256, the protocol is less efficient for a small number of tags (n between 0 and 350), but its performance is improved for higher n. The parabolic behaviour of the curve for L = 256 is not apparent in the figure, since occurs for higher values of n with respect to the simulation scope.

In figure, classic FSA is compared with Dynamic FSA, in which the frame length is changed dynamically in such a way to coincide always with the number of contending tags. When the estimation of the number of tags is exact ($\alpha = 1$), the DFSA protocol is able to significantly improve the performance of classic FSA. As we observe from the figure, the improvement is on the order of 200 time slots with respect to FSA with L = 128 and n up to 250. When n increases, the advantage of adopting DFSA instead of FSA becomes more and more relevant.

As an example of the binary tree algorithm, we consider a distributed binary tree (DBT) protocol that is self-adjusting, and that is directly managed by tags. It recalls the classic version of the binary tree traversal, in which, when a collision occurs, each client randomly chooses a binary value. This is the same principle at the basis of the binary selection phase in the EB-FSA protocol. In DBT, the reader sends its query and all tags randomly choose a binary value. Tags that chose 0 try transmission at the first time slot available, while the others try transmission at the following time slot. If a collision occurs, the same procedure used in EB-FSA is adopted. So, colliding tags randomly select another binary value, while all the others increment their counters by 1.

Such implementation of the BT protocol is independent of the bit coding of the tags IDs (that instead must be suitably chosen in the RFID standard BT protocol). Moreover, a fundamental role in RFID standard binary tree is played by the tag reader that must perform the splitting procedure based on the tags IDs. On the contrary, in DBT the tags are able to manage the protocol autonomously, without the reader's queries. This way, the tag reader is not required to store the status of forked queries to avoid travelling each edge more than once. On the other hand, a drawback of DBT is that it requires tags to perform some processing (as for generation of pseudo-random binary values), so it could be difficult to implement with passive tags.

Performance of the DBT protocol is very close to the theoretical expectation. DBT is able to improve significantly the performance of standard FSA and, for a number of tags up to 500, gives a moderate performance loss with respect to DFSA.

As a further assessment, we can compare the performance of the DBT protocol with that of EB-FSA. A simple example of application of the two protocols, where we consider the case of n = 5 tags. We observe from the figure that the only difference between the two protocols consists in the distribution of the initial values of the tag counters: in EB-FSA each tag must know the frame size (L = 5, in this case) and chooses its random value accordingly. In DBT, instead, each tag starts by choosing a binary value, without any knowledge on the frame size. However, even without needing any information on the number of tags, the DBT protocol is able to achieve the same performance as the EBFSA, in the considered example, since both protocols complete identification in 8 time slots. Moreover, contrary to EB-FSA, DBT does not require any estimation phase before identification.

On the other hand, it should be observed that the DBT protocol produces a higher number of collisions with respect to EB-FSA, due to the fact that the initial values chosen by the tags are only binary. In EB-FSA, instead, tags can select initial values in the range between 1 and nest. For these reasons, the DBT protocol is less efficient than EB-FSA under the power consumption viewpoint.

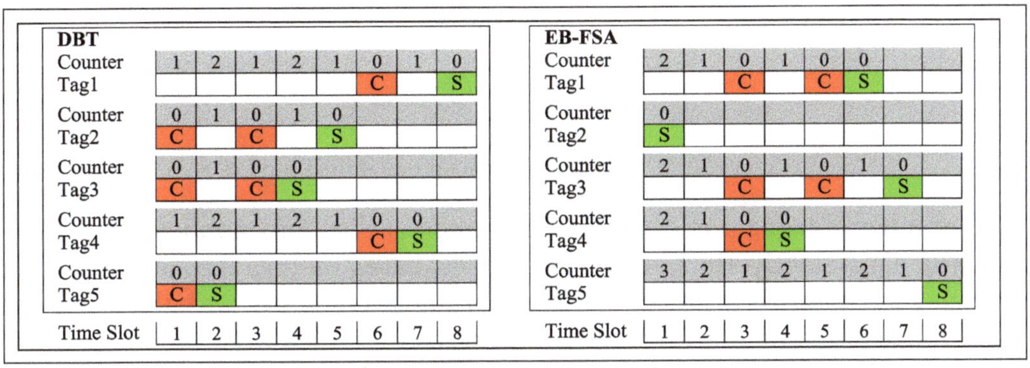

Example of application of DBT and EB-FSA protocols.

When the number of tags increases, the advantage of having many possible initial random values in EB-FSA becomes more and more relevant, yielding a performance improvement with respect to the DBT protocol. This is shown in figure above where DBT is compared with EBFSA through numerical simulations, for a number of tags up to 500.

As we observe from the figure, under the hypothesis of perfect estimation of the number of tags, the EB-FSA protocol outperforms the DBT one. However, we can take into account the number of time slots needed by the initial estimation phase of EB-FSA, I_{est}. The value of Iest has been found by considering, for the estimation phase, the same choice of the parameters, that is, $L_{est} = 64$, $P_{coll\text{-}th} = 0.7$ and $f_d = 4$. By considering the estimation phase, the performance gain achieved by EB-FSA becomes smaller and the two protocols have almost the same performance for a number of tags up to 300. So, the DBT protocol could still represent a valid choice, since it does not require the initial estimation phase, that has some drawbacks under the complexity viewpoint.

Another important aspect that must be taken into consideration is that, in the EB-FSA protocol, the estimation phase could be inaccurate, so the performance gain with respect to DBT could be further reduced. We consider two different cases, in which we suppose that the reader estimates a number of tags equal to 0.5 and 1.5 times the actual number. The results of numerical simulations of the EB-FSA protocol with inaccurate estimation, where they are compared with those of the DBT algorithm. We observe that the DBT protocol achieves almost the same performance as the EB-FSA in both the considered cases with inaccurate estimation. The overhead due to the initial estimation phase in EB-FSA has been also taken into account.

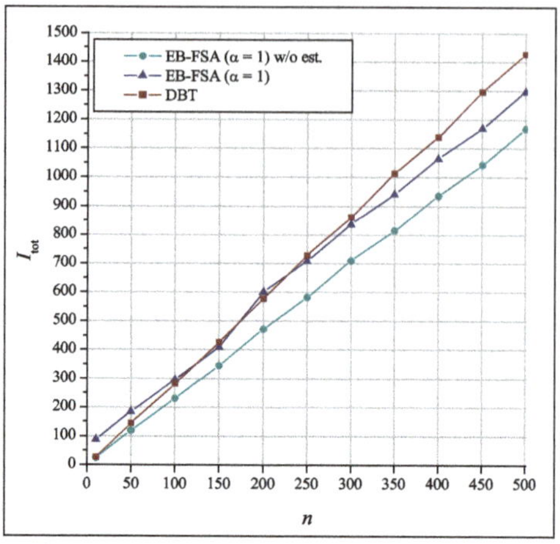

Comparison of DBT and EB-FSA with perfect estimation (α = 1).

So, for a number of tags up to 500 (that is of interest for many applications), the DBT protocol (or, equivalently, the binary tree protocol based on tags IDs) is able to guarantee a rather good collision arbitration. Its performance compares with that of optimized stochastic algorithms, as the EB-FSA. However, as we notice from the figure, the DBT curve has a higher slope with respect to those of EB-FSA and intersects them at n ≈ 300. This confirms that, when the number of tags increases, the advantage of EB-FSA becomes more and more relevant.

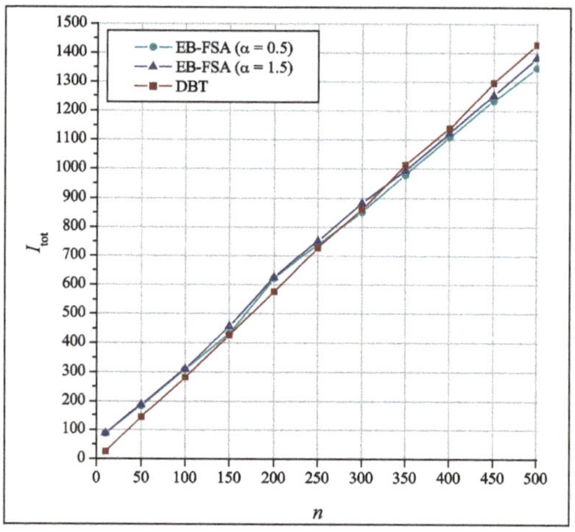

Comparison of DBT and EB-FSA with inaccurate estimation (α = 0.5, 1.5).

STOCHASTICAL MODEL AND PERFORMANCE ANALYSIS OF RFID

RFID (Radio Frequency Identification) can assign a unique digital identifier to each physical item, and provide an efficient, cheap and contactless method for gathering the information of the physical items to enable their automatic tracking and tracing. RFID technology serves as the back stone of the "Internet of Things", and is reviewed as a main enabler of the upcoming "Pervasive Computing". RFID systems have been widely adopted in quite a lot applications, ranged from "smart box" to world-wide logistics management systems.

A typical RFID system is consisted of some RFID tags, one or more RFID readers and the backend information system. Each RFID tag holds a unique identifier and is attached to a physical item. RFID reader is used to collect the identifiers stored in the RFID tags located in its vicinity and is often connected with the backend information system. During identification, The RFID reader asks the RFID tags to modulate their binary identifiers into signals and transmit these signals back to the reader through the air interface, which is a wireless communication channel for the RFID tag and reader to exchange information. Afterwards, The RFID reader sends the data gathered to the backend information system for further processing and dispatching to various applications.

One of the main issues that affect the universal deployment and application of the RFID system is the collisions occurred during RFID tag identification. The simultaneous modulated signals broadcasted by the RFID reader and transmitted from the RFID tags will interfere in the air interface, in which case, what the receivers can get is only a collision signal but no useful information. Collisions occurred in the RFID system can be categorized as the reader-reader, reader-tag and tag-tag collisions.

When two or more RFID readers try to broadcast messages through the air interface simultaneously, reader-reader collision occurs. Due to that they are often connected with the computer system and can be equipped with enough resource to monitor the air interface, RFID readers can detect the collision and coordinate with each other in advance. Reader-reader collision can be avoided and resolved completely with some deliberate designed protocols, such as the ColorWav and others.

Reader-tag collision can also be avoided easily by asking the RFID reader and RFID tags to broadcast message and transmit identifiers in different time, so that the signals broadcasted by the RFID reader and transmitted from RFID tags will never collide in the air interface and reader-tag collision will never occur.

When two or more RFID tags try to transmit their identifiers simultaneously through the air interface, tag-tag collision will occur. Due to the extreme constraints on computation, communication and energy supply put on RFID tags, especially that in passive RFID system, RFID tag can only get power supply through the reflection of the waveforms broadcasted by the RFID reader, tag-tag collision cannot be resolved easily using existed collision resolution methods, such as CDMA (Code Division Multi Access), FDMA (Frequency Division Multi Access), SDMA (Space Division Multi Access) and TDMA (Time Division Multi Access), proposed in other communication systems.

Proposed protocols for RFID tag collision resolution can be classified as the probabilistic frame slotted ALOHA based protocols, the deterministic splitting tree based protocols, and some hybrid protocols, such as the slotted tree protocol. The deterministic splitting tree based RFID tag collision resolution protocols suffer from scalability, and perform clumsily in resolving the collision caused by a large amount of RFID tags, while the hybrid protocols are seldom adopted in real RFID systems, so these collision resolution protocols.

Due to that in a frame, each RFID tag can only choose a slot to transmit its data, the RFID tag collision resolution process using the frame slotted ALOHA based protocols can be viewed as a binomial distribution process, the identification accuracy and the efficiency of the protocol depends seriously on the estimation and choice of some key parameters.

The Frame Slotted ALOHA based Protocols

The analysis of multiple-access protocols, we hold the following basic conventions in the discussions:

- Time is slotted, each time slot can be either a command slot for the RFID reader to broadcast a message or a data slot for the RFID tags to transmit their binary identifiers.
- One RFID tag group is interrogated in a data slot.
- The air interface is perfect, and no signal transmission error or lose occurs in it.
- The air interface is ternary, the interrogation of a tag group reveals the presence of zero, one and two or more RFID tags.

It is assumed that the air interface is instantaneous, not delayed, due to the limited distance between the RFID reader and RFID tags. However, short time intervals are needed for the RFID reader and tags to modulate and transfer a bit through the air interface. As other research work on RFID tag collision resolution protocols, it is assumed that RFID tags remain stable in a collision resolution cycle, neither newly arrived RFID tag enters nor does existed RFID tags leave the interrogation zone of the RFID reader during an identification cycle.

The Frame Slotted ALOHA based RFID Tag Collision Resolution Protocols

The ALOHA protocol for resolving the collision occurred in wireless communication was originally proposed by N. Abramson from the Hawaii University in the 1970s to enable the terminals distributed in isolated islands to exchange information with the mainframe computer system. In this protocol, each terminal can choose a time interval randomly to transmit its data to the mainframe, and if the time interval is occupied solely by the terminal, the data can be received by the mainframe successfully. Otherwise, if the time interval is occupied by two or more terminals and collapses, the data signals from these terminals will interfere, and collision will occur, and each terminal is acknowledged with the collision and will choose randomly another time interval afterward for retransmission. An improvement of the ALOHA protocol is the slotted ALOHA protocol, in which the time is slotted, each terminal can only choose randomly a whole time slot, start data transmission at the beginning of the time slot and finish transmission at the end of it. In such a way, a lot of collisions occurred in the wireless communication channel can be avoided.

Due to its simplicity and easiness of implementation, the slotted ALOHA protocol was introduced to the RFID system to resolve the tag-tag collisions, which are caused by that two or more RFID tags try to transmit their identifiers simultaneously through the air interface, and various frame slotted ALOHA based RFID tag collision resolution protocol have been suggested, such as the standard frame slotted ALOHA protocol, the dynamic frame slotted ALOHA protocol, the extended frame slotted ALOHA protocol and etc. Some frame slotted ALOHA protocols have been adopted as international or industrial standards. In all these protocols, the overall process that a RFID reader tries to collect the identifiers stored in the RFID tags within its vicinity is called a collision resolution cycle (or an identification cycle), which is consisted of a series of frames. And in each frame there is a command slot and a series of data slots.

In the command slot of a frame, the RFID reader broadcasts a command message to RFID tags in its interrogation zone to indicate the number of data slots (frame length), s, adopted in this frame. Each RFID tag, upon receiving this message and decoding the frame length s, randomly generates an integer i in the range 0..s-1, modulates its binary identifier into signals and transmits the signals back to the RFID reader in the i_{th} data slot through the air interface. Due to the constraint on computation put on RFID tags, s is often chosen as a integer mean of 2 and in the range [2,4,8,16,32,64,128,256].

Afterwards, the RFID reader gathers information contained in the data slots of the frame. If in a data slot, no RFID tag chooses to transmit its identifier, the data slot is called an idle slot. Else if in a data slot, there is one and only one RFID tag choosing to transmit its identifier, the data slot is called a success slot, the identifier is gathered and the RFID tag is identified successfully. Otherwise, two or more RFID tags choose to transmit their identifiers in the data slot, collision will occur and the data slot is called a collision slot, in such case, what the RFID reader can get is only a collision signal but no other useful information.

If after a frame, the desired identification accuracy of RFID tags specified by the application system is achieved, the current collision resolution cycle can be terminated, and the RFID reader will report the result to the backend information system for further processing and dispatching. Otherwise, another collision resolution frame is needed, and the same identification process is repeated.

The Stochastical Model for Collision Resolution of RFID Tags

The Binomial Distribution Model for RFID Tag Collision Resolution

In a collision resolution frame, due to that each RFID tag can only randomly choose one data slot to transmit its digital identifier this procedure can be viewed as a typical binomial distribution process.

Suppose that the population of RFID tags within the vicinity of the RFID reader is t and the frame length is s, the probability that n RFID tags choose a common data slot to transmit their identifiers can be calculated with the binomial distribution as,

$$B_{s,\frac{1}{t}} = \binom{t}{n}\left(\frac{1}{s}\right)^{n}\left(1-\frac{1}{s}\right)^{t-n}$$

where $\binom{t}{n} = \frac{n!}{t!(n-t)!}$. And the mathematical expects of this binomial distribution can be calculated with $E\left[\left[B_{t,\frac{1}{s}}(n)\right]\right] = s\binom{n}{t}\left(\frac{1}{s}\right)^{n}\left(1-\frac{1}{s}\right)^{t-n}$

Suppose that μ_r is a random variable representing the amount of data slots that r RFID tags chooses the data slot to respond, where r∈[0,1,..,t-1], the distribution of μ_r, according to, is

$$P(\mu_r = m) = \frac{\binom{s}{m}\prod_{k=0}^{m-1}\binom{t-k}{r}G(s-m,t-rm)}{s^{t}}$$

Where $G(M,m) = M^{m} + \sum_{k=1}^{\left\lfloor\frac{m}{r}\right\rfloor}\left\{(-1)^{k}\prod_{j=0}^{k-1}\left\{\binom{m-jr}{r}(M-j)\right\}(M-k)^{m-kr}\frac{1}{k!}\right\}$. The mathematical expects for the number of idle, success and collision data slots, $C_{0,t,s}$, $C_{1,t,s}$, and $C_{k,t,s}$ achieved in the frame can be calculated as,

$$C_{0,t,s} = s\left(1-\frac{1}{s}\right)^{t} \approx s e^{-\frac{t}{s}}$$

$$C_{1,t,s} = t\left(1-\frac{1}{s}\right)^{t-1} \approx t e^{-\frac{t}{s}}$$

$$C_{k,t,s} = s - C_{0,t,s} - C_{1,t,s} \approx S\left(1-\left(1+\frac{t}{s}\right)e^{-\frac{t}{s}}\right)$$

The values of $C_{0,t,s}$, $C_{1,t,s}$, and $C_{k,t,s}$ for the identification of different amount of RFID tags with fixed frame length s=256. It can been observed that for the collision resolution protocol with fixed frame length, as tag population increases, the amount of collision slots also increases rapidly and approaches to the frame length, and the amount of idle slots decreases rapidly and approaches to 0, while the amount of success slots, after reaches a maximum value, will also decrease rapidly and approach to 0 finally.

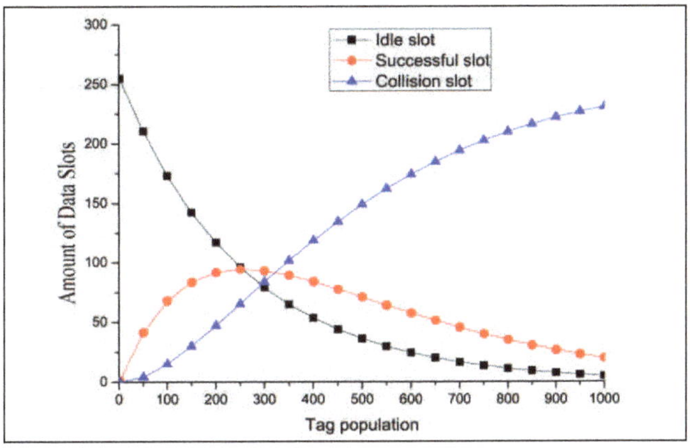

The values of $C_{0,t,s}$, $C_{1,t,s}$, and $C_{k,t,s}$ for different tag population and fixed frame length s=256.

Estimation of RFID Tag Population

In typical RFID applications, the population of RFID tags within the vicinity of the RFID reader is usually unknown in advance. For the frame slotted ALOHA based RFID tag collision resolution protocols, accurate estimation of the RFID tag population is a necessity for the calculation of the identification accuracy achieved after a frame and the termination of the current collision resolution cycle.

Proposed methods for the estimation of the RFID tag population adopted in the frame slotted ALOHA based protocols, according to the name of the proposer, can be categorized as the Vogt-1, the Zhen-1, the Cha-1, the Cha-2 and the Vogt-2 methods.

The Vogt-1 method refers to that after an identification frame, if the amounts of idle, successful and collision slots are a_0, a_1, and a_k, due to that a1 RFID tags respond in the successful slots and that every collision is occupied by the identifiers from at least 2 RFID tags, the overall population of RFID tags that respond in the frame can estimated as t directly with,

$$t = a_1 + 2a_k$$

The Zhen-1 method uses the probability that collision occurs in a data slot to estimate the overall population of RFID tags in the vicinity of the RFID reader. The probability that a RFID tag may collide with other RFID tags in a data slot can be calculated as,

$$C = \frac{P_{col}}{1 - P_{suc}} = 1 - \frac{\left(1 - \frac{1}{s}\right)^{t}}{1 - \frac{t}{s}\left(1 - \frac{1}{s}\right)t - 1}$$

where P_{col} and P_{suc} refer to probabilities that a data slot result in collision and success. According to the law of large number, the stable of value of C can be achieved when $t \to \infty$ and in such case, C can be calculated as C=0.418. So the overall population of RFID tags that transmits their identifiers in the frame can be estimated with:

Another method is that the overall population of RFID tags within the vicinity of the RFID reader should be estimated as $t=2.392a_k$, and this method is called the Cha-1.

According to the binomial distribution of the frame slotted ALOHA based RFID tag collision resolution protocols, the mathematical expects for the amounts of idle, successful and collision slots can be calculated with equation $P(\mu_r = m) = \frac{\binom{s}{m}\prod_{k=0}^{m-1}\binom{t-k}{r}G(s-m,t-rm)}{s^t}$, if after a frame, the actual number of collision slots is a_k, for the equation $a_k = s\left(1-\left(1+\frac{t}{s}\right)e^{-\frac{t}{s}}\right)$, for known s, the value of t can be calculated with the Newton or other literation methods.

This calculation is time consuming, so to ease the calculation, for the framel length s adopted in the protocol, we can calculated the value of $C_{k,t,s}$ for different t, and compare a_k with it to find the approximate RFID tag population t. That is, through searching t to minimize the $\left|a_k - C_{k,t,s}\right|$, the appropriate value of RFID tag population can be founded. This method is called Cha-2.

It should be noticed that the search range of t should be limited to $[a_1+2a_k, 2(a_1+2a_k)]$, for that (a_1+2a_k) is the low limit of the tag population, and numeric simulation shows that the chance is rare for actual RFID tag population exceeds $2(a_1+2a_k)$.

Accroding to the Chebyshev's inequality, the result for random experiment with random variable X is most likely results in the mathematical expects of X, and to resolve the collision caused by different number of RFID tags t with frame length s, the mathematical expects of $C_{0,t,s}, C_{1,t,s}$, and $C_{k,t,s}$ achieved in a frame can be calculated in advance. And if after the frame, if the actual amounts of idle, success and collision slots are a_0, a_1, and a_k, the population of RFID tags t can be calculated by minimizing the difference between the tuples $< a_0, a_1, a_k,>$. That is to find the RFID tag population t, which satisfies $< C_{0,t,s}, C_{1,t,s}, C_{k,t,s} >$. That is to find the RFID tag population t, which satisfies, $\mathrm{Min}(f) = \mathrm{Minimize}\begin{vmatrix} C_{0,t,s} - a_0 \\ C_{1,t,s} - a_1 \\ C_{k,t,s} - a_k \end{vmatrix}$

The search range of RFID tag population t can also be limited to the range $[(a_1+2a_k), 2(a_1+2a_k)]$.

These methods estimate the population of RFID tags within the interrogation zone of the RFID reader from different views, and the estimation accuracy achieved in each method needs to be examined with further numeric simulations.

The Calculation of the Optimal Frame Size

In the frame slotted ALOHA based RFID tag collision resolution protocols, once the population of RFID tags is known or can be estimated, the choice of the frame length adopted in the protocol affects the efficiency of the protocol and the latency of a collision resolution cycle. The choice of the optimal frame size should take into consideration of both the throughput of the protocol and the efficiency of RFID tag identification.

The throughput of the collision resolution protocol reflects the efficient use of the air interface, and is defined as,

$$e_{th} = \frac{c_{1,t,s}}{s} = \frac{t(1-1/s)^{t-1}}{s}$$

where t is the RFID tag population, and s refers to the frame length adopted.

For e_{th} to achieve its maximize value, we need to fix t, and let $\frac{de_{th}}{ds} = 0$. It can be found that when s=t, e_{th} is maxi minimized to be $\left(1-\frac{1}{t}\right)^{t-1}$. Similarly, according to the law of large number, the stable maximum value of e_{th} can be calculated as,

$$\max(e_{th}) = \lim_{t\to\infty}\left(1-\frac{1}{t}\right)^{t-1} = e^{-1} = 0.368$$

An alternative is to view this collision resolution process as a Poisson distribution process. The probability that t RFID tags transmit their identifiers back to the RFID reader in a time interval $[0, \tau]$ is in accordance with the Poisson distribution, and can be calculated with,

$$p(X(\tau)=k) = \frac{(\lambda\tau)^k}{k!}e^{-\lambda\tau}$$

where $\lambda = \frac{t}{s}$.

Due to that the frame slotted ALOHA based RFID tag collision resolution protocols divide an identification frame into a series of discrete data slots, and each slot can be viewed as one time unit, if only one RFID tag chooses the time unit to transmit its identifier, no collision occurs and the RFID tag is identified successfully, so the throughput of the frame slotted ALOHA based collision resolution protocol can viewed as $p(X(1)=1) = \lambda e^{-\lambda\tau}$, and when $\lambda = 1$, p is maximized to be 0.368. This also verifies that when s=t, the throughput of the protocol is maximized.

The throughputs achieved by the frame slotted ALOHA based RFID tag collision resolution protocols with different frame length in the identification of different amount of RFID tags.

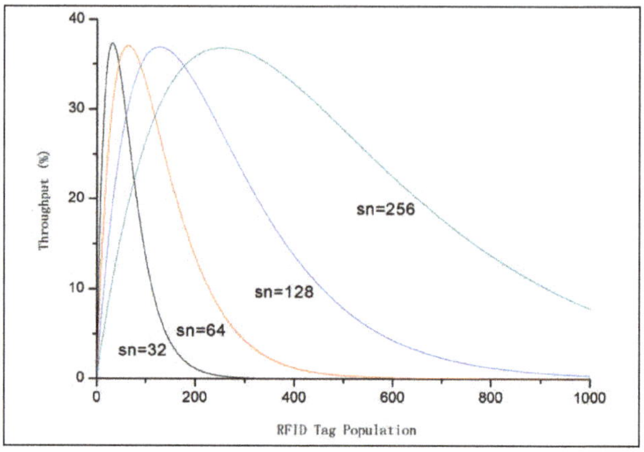

The Throughputs Achieved by the Frame Slotted ALOHA Based RFID Tag Collision Protocols with Different Frame Length.

In the research of frame slotted ALOHA based RFID tag collision resolution protocols, throughput is often used to determine the optimal frame size adopted in the protocol. But we think that although throughput in an important issue to measure the efficient use of the communication channel, another key factor should be taken into consideration for the calculation of the optimal frame length is to consider the performance of the collision resolution protocol using the identification ratio of RFID tags achieved in a frame, which can be defined as $e_{s,t}$ and calculated with:

$$e_{s,t} = \frac{c_{1,t,s}}{t} = (1 - /s)^{t-1}$$

To find the maximum value of the identification ratio $e_{s,t}$ we also need to fix s, calculate $\frac{de_{s,t}}{ds}$, and let $\frac{de_{s,t}}{ds} = 0$, we get:

$$\frac{de_{s,t}}{ds} = \left(1 - \frac{1}{s}\right)^{t-1} + t\left(1 - \frac{1}{s}\right)^{t-1} \text{In}\left(1 - \frac{1}{s}\right)$$
$$= \left(1 - \frac{1}{s}\right)^{t-1} \left(1 + t\text{In}\left(1 - \frac{1}{s}\right)\right) = 0$$

and then $s = \frac{1}{1 - e^{-\frac{1}{t}}} = 1 + \frac{1}{e^{\frac{1}{t}-1}}$.

According to the discussion presented above for the identification of different amount of RFID tags in the vicinity of the RFID reader, the corresponding optimum frame length for the frame slotted ALOHA based protocol to achieve best identification ratio can be calculated. Due to that the frame legnth adopted in the protocol can only be chosen in the range of [2,4,8,16,32,64,128,256], for the identification of t RFID tags, the appropriate frame length s should satisfy:

- $e_{s,t} > 2e_{\frac{s}{2},t}$, which means that the amount of RFID tags identified in a frame with frame length s should be more than that identified in two frames with frame length $\frac{s}{2}$.
- $2e_{s,t} > e_{2s,t}$, means that the amount of RFID tags identified in two frames with frame length s should be more than that identified in a frame with frame length 2s.

Collision Resolution Process based on the Markov Chain

Suppose that in a collision resolution cycle of the frame slotted ALOHA based RFID tags collision resolution protocol, after the i_{th} frames, the amount of identified RFID tags is f(i), then after the next frame, the amount of RFID tags identified should be f(i+1)=f(i)+-t_i, where t_i is the amount of RFID tags which are newly identified in the frame i+1 but have not been identified in the previous frames. This specifies that the amount of RFID tags identified after frame i+1 depends solely on the amount of RFID tags identified after frame i, and this process can be viewed as a homogenous Markov chain.

The Markov chain is often defined using the transition matrix to specify the probability that a state changes to another. The elements of this Markov chain transition matrix for the identification of t RFID tags using the frame slotted ALOHA protocols can be calculated with,

$$q_{i,j} = \begin{cases} 0 & (j<i) \\ \sum_{r=0}^{i} P(\mu_1 = r)\dfrac{\binom{i}{r}}{\binom{t}{r}} & (j=i) \\ \sum_{r=j-i}^{i} P(\mu_1 = r)\dfrac{\binom{t-i}{j-i}\binom{i}{r-j+i}}{\binom{t}{r}} & (j>i) \end{cases}$$

Each elment $q_{i,j}$ specifies the probability that the amount of identified RFID tags changes from i to j after a frame.

The first situation specified in equation above will never occur due to that it is impossible that after a new frame the total amount of RFID tags identified is less than that identified before the frame.

The second situation specifies that the amount of RFID tags newly identified in the frame is 0, which means that all RFID tags which are identified without collision in this frame have been identified in the previous frames, and the current collision resolution cycle should be terminated because that the probability that new RFID tags can be detected in the following frames is also 0. The coefficients for such transition can be calculated with the equation,

$$q_{ii} = 1 - \sum_{j=0,j\neq i}^{t} q_{ij} = 1 - \sum_{j=1,i+1}^{t} q_{ij}.$$

For the third situation, of all t-i RFID tags not identified in the previous frame by the RFID reader, j-i RFID tags choose the success data slot to respond and are identified newly in the frame.

The values for elements in the first row of the transition matrix specifies the initial state of a collision resolution cycle, and should be set to $q_{oj} = \{1,0,...0\}$.

The Markov chain and corresponding transition matrix specifies the condition that a collision resolution cycle can terminate, and can be used to calculate the number of frames needed for the identification of t RFID tags.

The Employment of Multiple RFID Readers

Usually in a dense RFID tag environment, multiple RFID readers are deployed to facilitate the RFID tag identification cycle. Suppose that there are n readers deployed, and each reader resolves the RFID tag collision independently and reader-reader collision is resolved. For the overall identification accuracy α required by the application system, the accu-racy 紘 which each RFID reader should achieve can be calculated as,

$$(1-\gamma)^n \leq 1-\alpha$$

and we have

$$\gamma \geq 1-\sqrt[n]{1-\alpha}$$

Table: shows that if overall identification accuracy required by the application systems is 99.0%, and multiple readers are deployed, the identification accuracy which each RFID reader should achieve. From table below, we can see that the deployment of multiple RFID reader decreases the accuracy requirement for each reader significantly, which will in return, facilitate the identification cycle greatly.

Table: Identification Accuracy for Each RFID Reader.

Number of RFID readers deployed	Identification accuracy required for each RFID reader
1	99.0%

2	90.0%
3	78.5%
4	68.4%
5	60.2%

Numeric Simulation and Result Analysis

The Numeric Simulation Environment

In the simulation, 100 randomly generated data sets are used, in each data set, there are 1000 randomly generated binary strings, and each of which represents the binary identifier of a RFID tag encoded with SGTIN-96 schema. The standard frame slotted ALOHA based RFID tag collision resolution protocols with different frame length are implemented and simulated with the C# programming language in Microsoft Visual Studio.

.NET 2005 for the measurements of their performances in resolving the collision caused by different amount of RFID tags contained in each data set, the results are recorded and averaged with the 100 data sets.

The Accuracy of RFID Tag Population Estimations

To find the accuracy for RFID tag population estimation of various methods simulations are performed, in which the frame size of the frame slotted ALOHA protocol is fixed to 256. The accuracies of the RFID tag population estimation methods are measured with the mathematical means and variances of their estimation error ratios achieved in the simulations.

The mathematical means of the estimation error ratios for a RFID tag population estimation method is calculated as,

$$\mu_{err} = \frac{\sum_{i=1}^{R} \frac{\hat{t}-t}{t}}{R} = \frac{\sum_{t=1}^{R} \hat{t}}{Rt} - 1$$

And the mathematical variance of the estimation error ratio for a RFID tag population estimation method is calculated as,

$$\sigma_{err} = \sqrt{\frac{\sum_{i=1}^{R} \left(\frac{\hat{t}-t}{t} - \mu_{err} \right)^2}{R-1}}$$

where t and $\hat{t}$ represent the actual and estimated RFID tag populations. R is the number of data sets used the simulation, and in this example, and is fixed to 100.

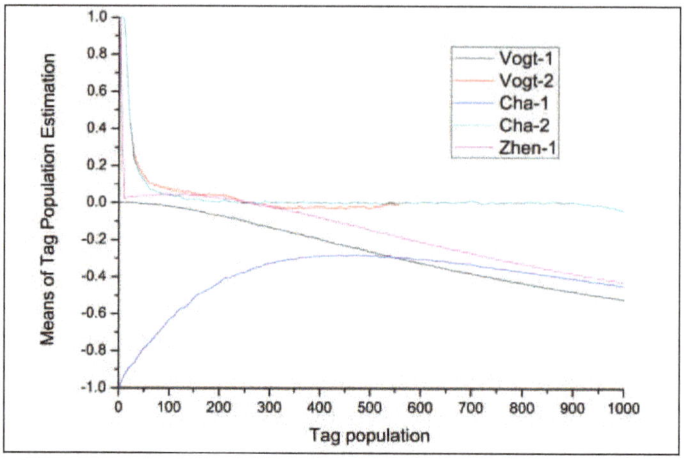

Mathematical Means of the Tag Population Estimation Error Ratios.

Figure above shows the mathematical means of the RFID tag population estimation error ratios of the Vogt-1, Vogt-2, Cha-1, Cha-2 and Zhen-1 methods, from which it can be concluded that the Vogt-2 method performs better than other methods with stable means of error ratios around 0.

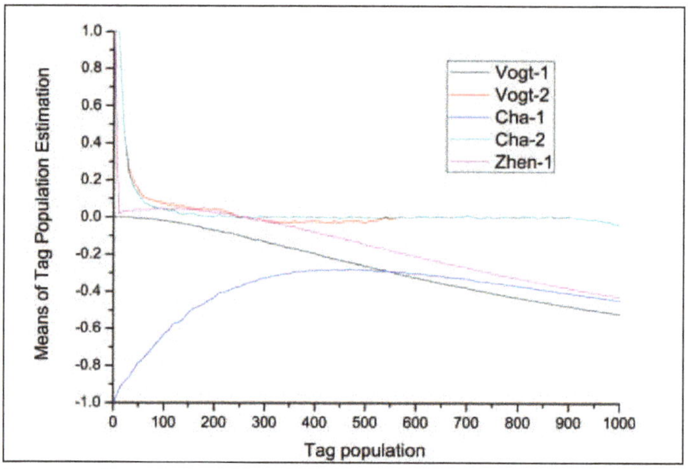

Mathematical Variances of the Tag Population Estimation Error Ratios.

Figure above shows the mathematical variances of the estimation error ratios of these methods, from which it can also been seen that although the variance of the tag population estimation ratios for Vogt-2 is the greatest, but is still within a satisfactory range.

For the tag population estimation using Vogt-2 and Cha-2, as we have discussed, search on tag population t is needed to find the minimal value of the evaluation function, and the search can be limited in the range $[a_1 + 2a_k, 2(a_1 + 2a_k)]$. In the simulations, we examine the probability that the actual tag population is in the range, and the result is shown in figure. From these simulations, we have observed that if the tag population is less than 3.2 times of the frame size, this upper limit $2(a_1 + 2a_k)$ has never been exceeded.

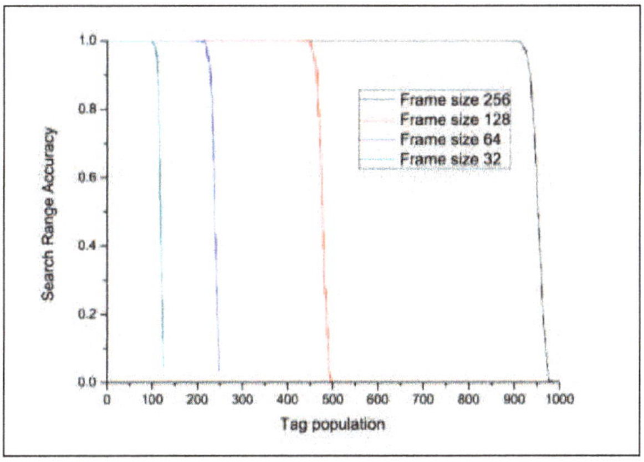

The Probability that the Tag Population is within the Range.

The Efficiencies of the Frame Slotted ALOHA Protocol And Analysis

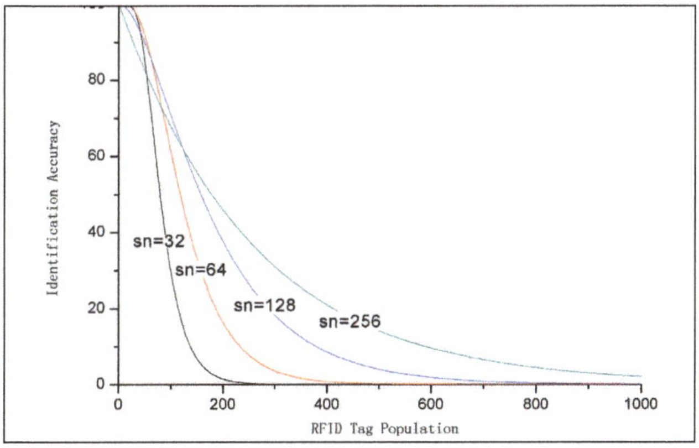

The Efficiencies of Frame Slotted ALOHA Protocols with Different Frame Length.

Figure above shows the efficiency of the frame slotted ALOHA protocols with different frame length in resolving the collision caused by different amount of RFID tags. For the convenience of comparison, the protocol with frame length s is performed 256/s frames, for example, the collison resolution protocol with frame size 16 is performed 16 frames. The efficiency is defined as the identification ratio of RFID tags in these frames, calculated with the number of RFID tags actually identified divided by the actual number of RFID tags. It can be observed that as the population of RFID tags increase, the efficiency of frame slotted ALOHA protocol decreases rapidly.

According to the calculation and simulation, the optimal frame length which the frame slotted ALOHA protocol should adopt in resolving the collision caused by different number of RFID tags is shown in table.

Table: Optimal Frame Length for the Identification of Different Number of RFID tags.

RFID Tag Population	1-14	15-30	31-61	62-124	124~
Optimal Frame Length	16	32	64	128	256

Simulation and Analysis of the Identification Process

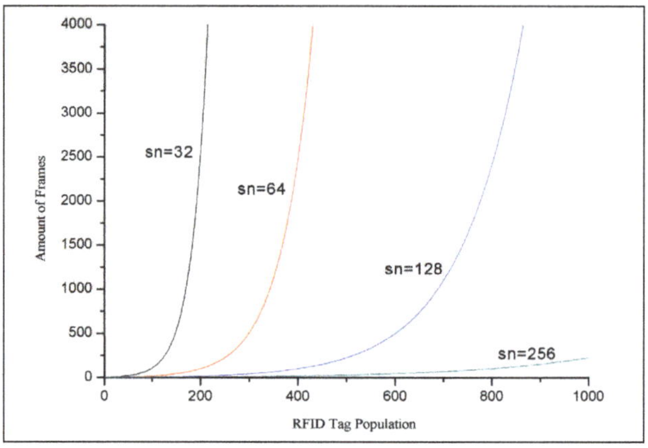

Amount of Frames Needed for the Frame Slotted ALOHA Protocol with Different Frame Length.

Figure above shows the amount of frames needed in resolving the collision caused by different amount of RFID tags using the frame slotted ALOHA based protocols with different frame length. It can be seen that as the RFID tag population increases, the amount of frames needed by these protocol will increase rapidly in exponential order.

LOW FREQUENCY RFID

RF design is often perceived as being very difficult and hard to understand. This application report guides you from the principles of the low-frequency (LF) RF communication to the right passive components and finally to a good layout.

The transmit frequency of the TI low-frequency RF products is 134.2 kHz. The communication range for passive devices (powered by the magnetic field of a RFID reader) is typically up to 10 cm.

Low-frequency RF Communication

Texas Instruments offers different types of low-frequency RFID devices. All passive TI RFID devices require a charge phase in which a capacitor is charged (TI half duplex (HDX) principle). Read-only devices then send their information (typically 96 bits). Read/write devices need to receive a command to perform a desired action and answer to the downlink command. Active 3D devices (for example, for passive entry or passive start devices) use the downlink channel to receive a command via LF and wake a microcontroller if the right wake pattern is received. The uplink is usually transmitted via UHF.

This application report describes the complete communication that is valid for all types of transponders.

Low-frequency RF communication can be used for short distances even in challenging environments. The direction of the antenna matters. Always use the antenna in its preferred direction. This should be kept in mind while designing a RFID system. Figure below shows how to position a stick antenna near an RFID base station antenna.

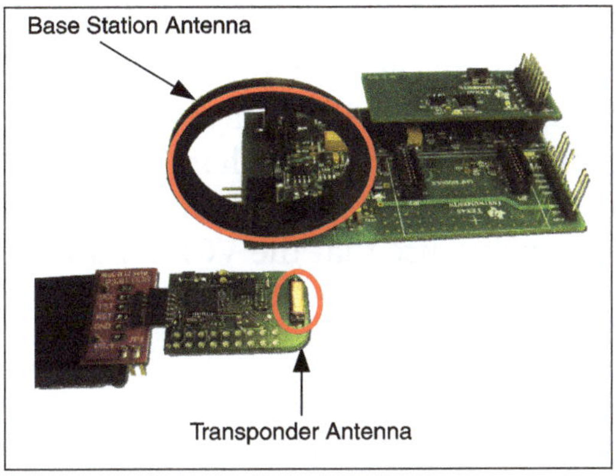

Correct Placement of a RFID Stick Antenna Near a Base Station.

Downlink - From the RFID Base Station to the RFID Transponder

An RFID communication is always initiated by the base station. The RFID transponder usually has no local voltage supply and needs the LF field, which is sent by the base station, to function.

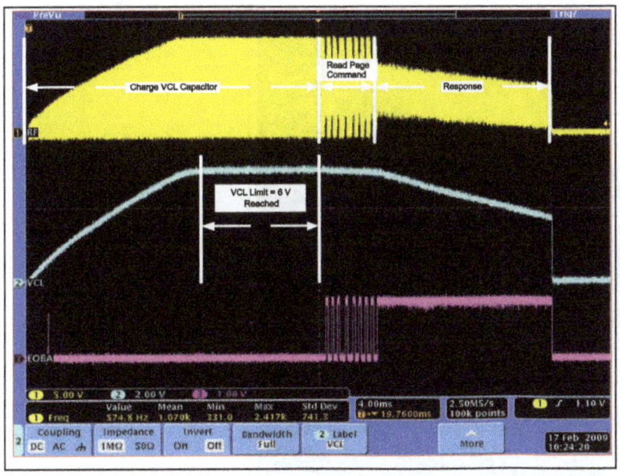

Read Page Command with Response (4 ms/div).

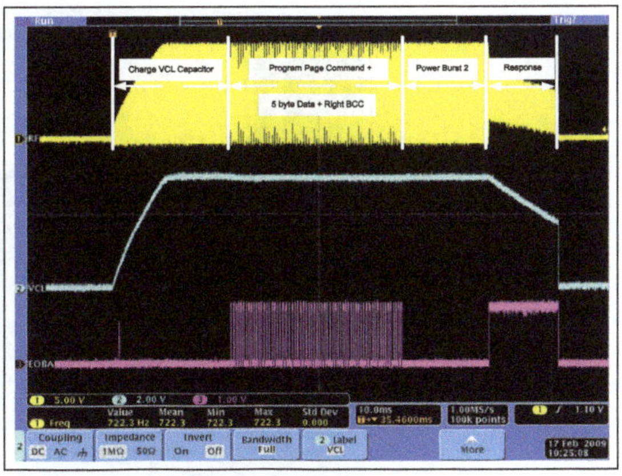

Program Page Command With Correct BCC and Response (10 ms/div).

A typical communication consists of a charge phase of 25 ms to 50 ms, which is followed by the command phase. A read and a program command are shown in figures respectively. In both cases the transponder sends its response after it detects a loss of the LF field.

All signals are measured at the transponder and have the following meaning:

Table: Signal Overview of Scope Print Outs.

NO.	NAME	COLOR	UNIT	MEANING
1	RF	Yellow	5 V/div	RF voltage at the transponder antenna
2	VCL	Blue	2 V/div	Voltage at the VCL charge capacitor
3	EOB	Magenta	1 V/div	End of burst detector output of transponder

The RF voltage is the voltage that is generated by the magnetic field of the RFID base station at the transponder antenna. The RF voltage is limited by the internal RF limiter of the transponder. The RF voltage is used to charge the VCL capacitor (typically 220 nF); VCL is the operating voltage for the transponder and is limited to approximately 6 V. For proper operation, make sure that the VCL limit is reached before the end of charging. The required charging time is a function of distance, antenna size, inductance, Q factor, and base station field strength.

After the charge phase, data is transmitted using On/Off Keying. The transponder decodes the received data. The EOB output of the transponder shows the detection of the bits. If the time between two rising edges of EOB is longer than t_{bitH} a high bit is detected; otherwise, a low bit is detected.

For a program page command, a 2-byte BCC is transmitted together with the data. Additionally, a second power burst is needed for programming the EEPROM. If the received BCC (needed for program and lock commands) is wrong or if the VCL level does not reach the VCL checker limit, the transponder performs a discharge without responding.

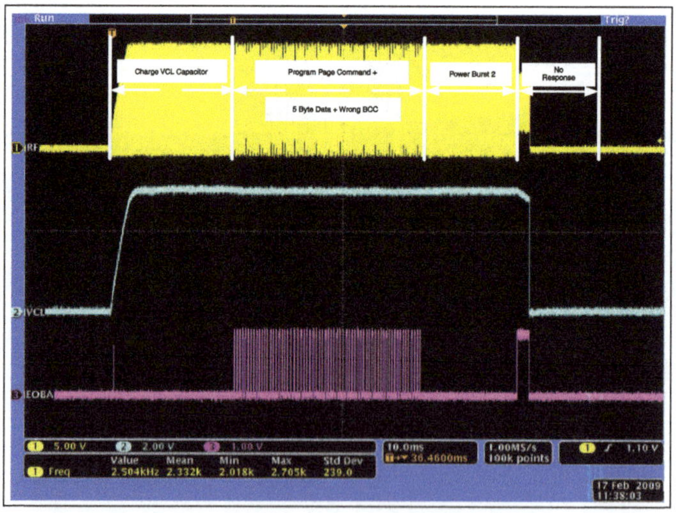

Program Page with Incorrect BCC and No Response (10 ms/div).

The downlink uses On/Off Keying for transmitting the data from the base station to the transponder. The base station switches the 134.2 kHz LF field on and off to modulate the data.

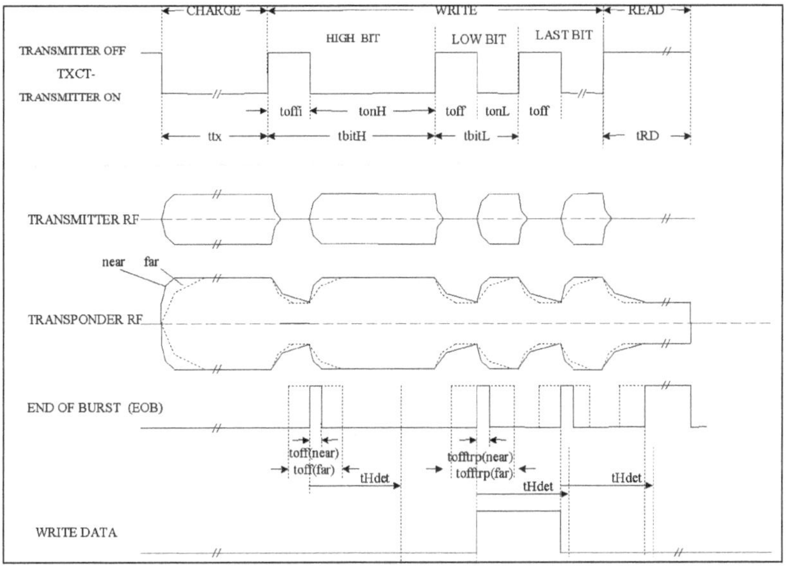

PPM Timing for a Read Command.

The bits are modulated with Pulse Position Modulation (PPM). Figure above shows the PPM timing for a read page command; Figure below shows the timing for a program page command. The program page command needs one additional transmitter off phase and a power phase. For TMS37157 the recommend timings are shown in table.

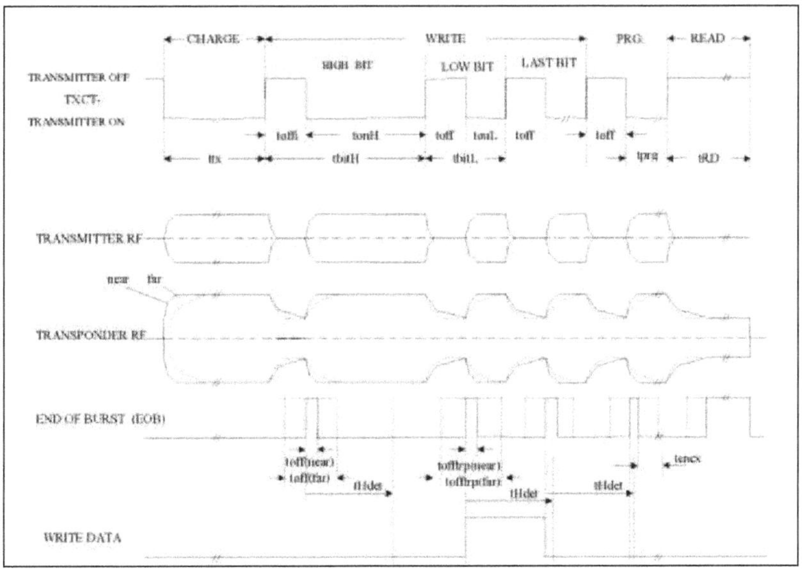

PPM Timing for a Program Command.

Table: Example Timings for the TMS37157 PaLFI Device.

SIGN	TIME	UNIT	NODE
t_{tx}	25	ms	Charge time to charge VCL capacitor
t_{offi}	170	μs	Initial off time

t_{off}	170	µs	Standard off time
t_{onH}	350	µs	Transmitter on time for a high bit
t_{bitH}	520	µs	Period for a high bit
t_{onL}	230	µs	Transmiter on time for a low bit
t_{bitL}	400	µs	Period for a low bit
t_{prg}	15	ms	Program time necessary for programming the EEPROM
t_{Hdet}	510	µs	High bit period must be 210 µs (minimum) to 1730 µs (maximum)

Uplink - From the RFID Transponder to the RFID Base Station

After reception of a valid command, the transponder processes the data and sends a response. The response is always 96 bits long and has the structure.

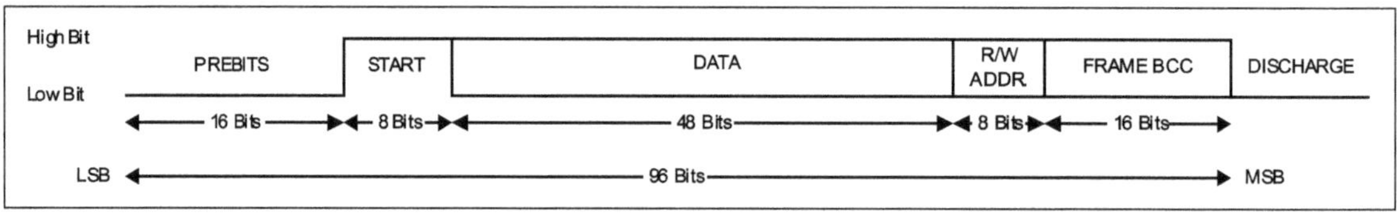

Response Format for a Transponder LF Command.

One high bit is represented by 16 sinus oscillations with a frequency of 123.7 kHz (length: 129.3 µs). One low bit is represented by 16 sinus oscillations with a frequency of 134.7 kHz (length: 128.3 µs).

Table: Description of the different fields of a transponder response.

FIELD	LENGTH	NODE
PREBITS	16 bits	Always 16 low bits - 0x00 00
START	8 bits	Depends on type of transponder - TMS37157 : 0x7E
DATA	48 bits	Depends on the performed command
R/W ADDRESS	8 bits	Depends on the performed command
FRAME BCC	16 bits	Calculated from START, DATA, and R/W ADDRESS

The response is always 96 bits long, resulting in a maximum duration for the response of approximately 12.2 ms. After transmitting the data, the transponder discharges the VCL capacitor. If the transponder receives an invalid command, it shows only the 16 pre-bits followed by discharging the VCL capacitor. If the transponder receives an invalid command, it shows only the 16 pre-bits followed by discharging the VCL capacitor.

How to Choose the Right Passive Components?

A Texas Instruments RFID low-frequency transponder system needs only a few passive components. This application report describes the components for 1D systems using the TMS37157. The application circuit for the transponder function of the TMS37157. The components and recommended values are listed in table.

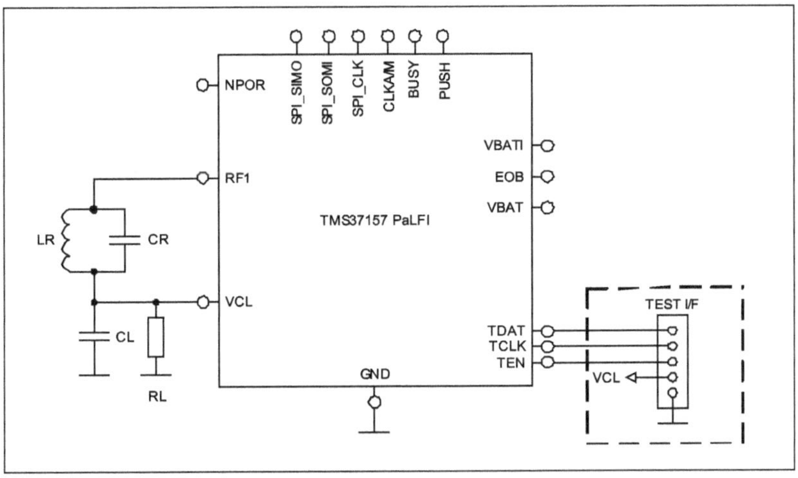

Transponder Only Application Circuit TMS37157.

Table: Recommended components for rfid transponders based on the TMS37157.

SYMBOL	FUNCTION	TYPE	VALUE
LR	Antenna	Ferrite stick antenna	2.66 mH
CR	Resonance capacitor	NPO	470 pF
CL	VCL charge capacitor	X7R	220 nF
RL	Termination resistor	Arbitrary	1 MΩ

Charge Capacitor CL and Resonance Capacitor CR

A half duplex (HDX) RFID system requires a charge capacitor to store energy for phases during which no magnetic field from the RFID base station is present. Texas Instruments recommends using a 220-nF surface mount capacitor, X7R type, with a maximum operating voltage of at least 16 V.

The resonance circuit of a RFID circuit is determined by the antenna and the resonance capacitor. Therefore, a resonance capacitor with a low variation in capacity and a high quality factor is required. Texas Instruments strongly recommends using a 470-pF ±2% surface-mount capacitor, NPO (CGO) type, with a maximum operating voltage of at least 20 V.

Only NPO (CGO) type capacitors have a high enough quality factor to ensure a properly working RFID system. Using another type of capacitor will likely result in a non-working system.

How to Measure the Quality Factor and Capacitance of a Capacitor?

To measure the quality factor and the capacitance of a capacitor, an impedance analyzer is needed. Figure shows the measurement performed with a Wayne Kerr impedance analyzer. Attach the capacitor between the two measurement pins of the impedance analyzer. The measurement frequency is 134.2 kHz (the nominal frequency of LF RFID).

Be aware that the quality measurement is not precise for such high values. The quality should be above 1000.

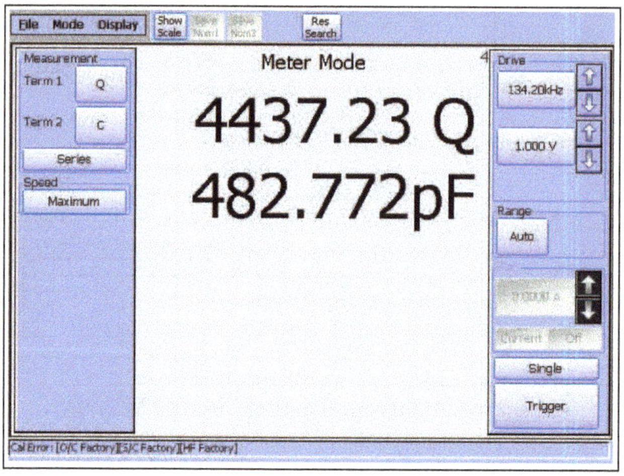

Quality Factor and Capacity of a NPO Capacitor.

Antenna

The quality factor of a NPO type resonance capacitor is normally above 1000. Thus, one of the most importing factors for the overall quality factor of the resonance circuit is the antenna. In combination with the TMS37157, Texas Instruments recommends using a ferrite stick antenna with an inductivity of 2.66 mH ±2.8%. To ensure an overall system quality of at least 30 (required for proper operation of the TMS37157) the quality factor of the antenna should be between 40 and 75.

How to Measure the Quality Factor and Inductivity of an Antenna?

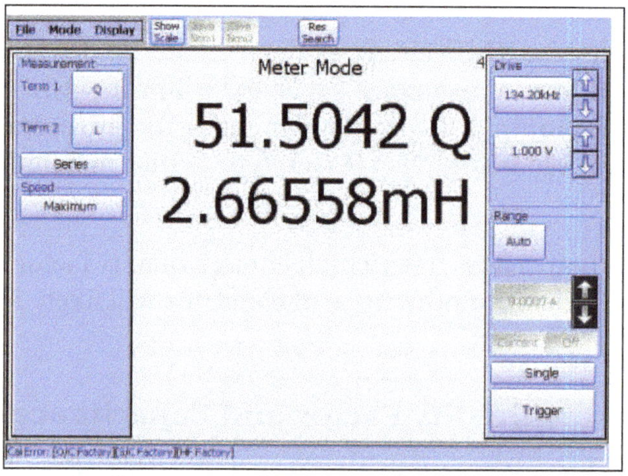

Quality Factor and Inductivity of an Antenna.

To measure the quality factor and the inductivity of an antenna, an impedance analyzer is needed. Figure below shows the measurement performed with a Wayne Kerr impedance analyzer. Attach the antenna between the two measurement pins of the impedance analyzer. The measurement frequency is 134.2 kHz (the nominal frequency of LF RFID).

Resonance Frequency and System Quality

How to Calculate the Resonance Frequency of a Resonance Circuit?

The resonance frequency of a resonance circuit can be calculated with equation:

$$f_{res} = \frac{1}{2\pi\sqrt{C_{res} \cdot L_{antenna}}}$$

Where:

- f_{res}: Resonance frequency of the resonance circuit.
- C_{res}: Resonance capacitor.
- $L_{antenna}$: Inductivity of the antenna.

The resonance frequency of the components from figures is 140.44 kHz. The TMS37157 has an additional internal parasitic capacitance of approximately 30 pF, which must be added to C_{res}. In addition, the TMS37157 has an integrated trimming function that offers the possibility to switch on additional resonance capacitors to trim f_{res} to 134.2 kHz.

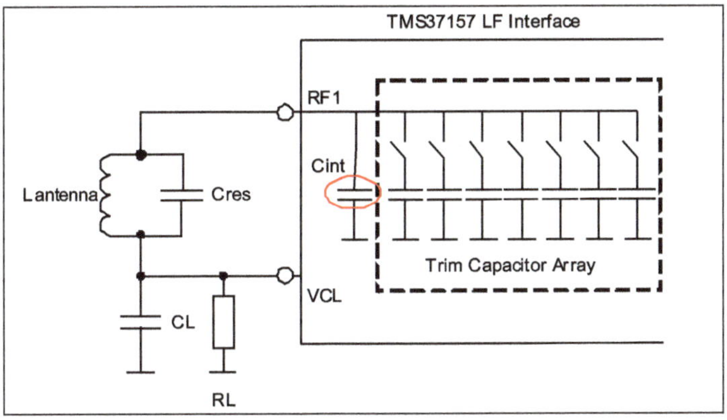

Internal Capacitances of TMS37157.

How to Measure the Quality Factor and Resonance Frequency of an RF Circuit?

For measurement of the quality and resonance frequency of an RF circuit, a spectrum analyzer is needed. Connect a send coil to the output of the spectrum analyzer and set the output power to minimum. For Texas Instruments low-frequency systems, choose a center frequency of 134.2 kHz. Connect another cable to one antenna pin and ground, the other end to the input of the spectrum analyzer. Be aware that high voltages can appear at the antenna, which can damage the input of the spectrum analyzer. It is recommended to place the RF circuit away from the send coil initially (20 cm or more), and then close the distance until a superelevation at the spectrum analyzer can be seen. The peak determines the resonance frequency. The quality factor can be calculated from the resonance frequency and the 3-dB bandwidth. The 3-dB bandwidth is the difference between the two frequencies where the power is half (3 dB lower) as in resonance. The quality factor can be calculated with equation:

$$Q = \frac{f_{res}}{B_{3dB}}$$

Where:

- Q: Quality factor of the resonance circuit.
- f_{res}: Resonance frequency of the resonance circuit.
- B_{3dB}: 3-dB bandwidth.

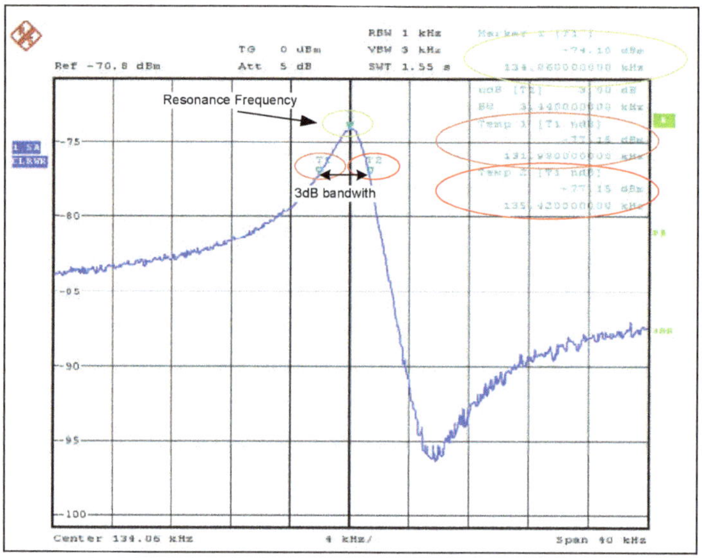

Resonance Curve of an RFID System.

The quality factor of the RFID system measured in figure is calculated with equation:

$$Q = \frac{f_{res}}{B_{3dB}} = \frac{134.06kHz}{3.440kHz} = 39$$

Influence of Resonance Frequency and System Quality on an RFID System

The TMS37157 requires a quality factor of at least 30 to work properly. The quality factor of the resonance circuit is determined by different factors:

- Antenna should have a high Q (52 to 60).

- Resonance capacitor must be NPO (CGO) type.
- In the PCB design, no metal layer should be underneath or around the antenna.
- No signal routing under the antenna.
- Keep PCB traces to the antenna short, straight, and thin.
- Keep antenna pads as small as possible.
- Quality factor and resonance frequency change with temperature.

An RFID system is described by the parts of the resonance circuit (antenna, capacitor) but also by the RFID chip and the other active and passive parts mounted on the PCB. The last step during production of an RFID system is to trim the circuit. Because of component tolerances, every RFID system must be trimmed—this can be easily achieved with the integrated trim function of the TMS37157.

A low quality factor and untrimmed resonance circuit decrease the performance of the whole RFID system. Inadequate energy is induced into the RF circuit of the transponder, resulting in a low voltage at the antenna (RF pin) and a poorly charged VCL capacitor. Metal under the resonance circuit detunes the resonance circuit and lowers the quality factor immensely; this should be considered while designing a mount-on-metal system.

Recommended PCB Layout

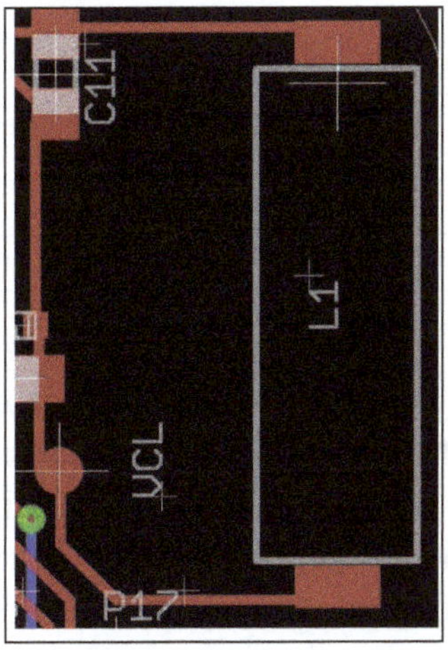

Recommended PCB Layout for an Antenna.

For a good PCB layout, keep in mind the factors which determine the system quality. An example for a recommended PCB layout is shown in figure where the L1 antenna and the C11 resonance capacitor are shown. The PCB traces to the antenna are straight, thin, and short and are not routed under the antenna. The antenna pads are as small as possible, and no metal layer is under or around the antenna.

HIGH FREQUENCY RFID

The High Frequency (HF) band is the frequency range from 3 to 30 MHz. RFID systems that operate in this frequency band are called High Frequency (HF) RFID Systems. HF RFID tags/readers usually operate at frequencies from 1.75 MHz to 13.56 MHz. NFC (Near Field Communications) is a sub-set of RFID and also operates in the high frequency (HF) band.

HF RFID tags are based on inductive coupling technology which means that the RFID reader induces a current in the RFID tag which then powers the RFID tag and sends out some basic identification information to the RFID reader. These tags are usually passive and do not require a power source or battery. They have anti-collision capabilities that allow a single reader to read multiple tags simultaneously. The range of these systems is usually under 1 meter (larger than LF).

High Frequency tags also work pretty well with/on metal objects and items having medium to high water content. HF RFID systems are used in a wide variety of applications including ticketing, payments, tracking library books, patient flow tracking and general data transfer applications.

Table: HF RFID Tag Properties.

Tag Type	Passive
Frequency	13.56 MHz (1.75 MHz to 13.56 MHz)
Read Rage	10 cm to 1 meter
Read Speed	Slow/Medium
Read Multiple Tags Simultaneously	Yes
Works with Metal	Yes
Works with Liquids	Yes
Cost	Low

Radio frequency identification (RFID) has received significant attention to serve as a means to digitally enhance the working or living environment. The popularity of using RFID derives from the fact that it provides explicit identification of objects at low cost. For over a decade, researchers have made use of the distinct information provided by RFID to create smart spaces. However, research on harnessing detailed location information from RFID tags is limited even though information about an object's location is of great importance for many applications.

Currently, the level of local resolution is limited to the places covered by RFID reader infrastructure. The grid of reader deployment determines the granularity of location information retrieved from an RFID tag where the reader's physical location is assigned to any tag within the range of that interrogator. Consequently, the detection of a RFID tag provides only limited local resolution – in close vicinity to a specific reader.

The exact physical location of the tag remains unknown. Currently, high frequency (HF) RFID readers neither resolve the distance nor the angle at which a tag is detected. In addition, the RFID infrastructure is often minimal (grid of interrogators) in order to keep the system costs low. A large grid size of reader deployment minimizes the degree at which the location of an object is resolved. This limitation is caused by the reader field that is usually non-directional to cover the largest area possible.

Passive HF tags operate on magnetic induction at 13.56MHz as opposed to ultra high frequency (UHF) tags that operate on the emission of electro-magnetic waves at 860-960MHz. Passive HF tags are widely used to equip items because of their small form factor, low cost, and resistance against interference (as opposed to UHF tags that show significant vulnerability to interferences).

Electro-magnetic sensing at radio frequency (RF) offers capabilities to determine the distance and direction of an object in relation to a reference point. The major techniques to derive these properties are time-of-arrival, time difference of arrival, angle-of-arrival, and received signal strength. Based on the distance measurement, the location is estimated through triangulation, which is done either through lateration that uses multiple distance measurements between reference points and the object or angulation, which measures angle or bearing relative to reference points. Research on location systems that are based on radio frequency is highly advanced. Other research has focused on mapping and localization through RFID tags that are distributed in the environment. In their research, mobile RFID readers are mounted on robots, which navigate according to the local information retrieved through the detection of embedded tags.

The techniques for RF distance-sensing described above apply for systems that operate on the emission of electro-magnetic waves such as RFID at UHF. In contrast, HF RFID systems operate on magnetic induction. This requires a different approach to resolve a tag's distance from a reader, because current techniques for distance-sensing are not applicable.

Sensing Principles

In inductively coupled systems the reader antenna generates a strong, high frequency electro-magnetic field to power the RFID tag and to serve as medium for data transmission. The electro-magnetic field penetrates the cross-section of the reader antenna coil and to some extent the antenna coil of the tag in close vicinity. By inductance, a voltage u_{tag} is generated at the tag's antenna. This inductively coupled system can be treated as a transformer-type coupling with a primary and a secondary coil as long as the tag remains in the near field – approximately up to 3.5m ($0.16 \cdot \lambda_{13.56MHz}$) from the antenna (idem). A tag in the interrogation field draws energy from the magnetic field generated by the reader. This energy consumption results in a voltage change at the reader's antenna due to mutual inductance. A tag transmits data by switching a load resistor on and off thereby influencing the transformed transponder impedance. This change in impedance causes an amplitude modulation of the voltage reader at the reader's antenna coil. The intensity of the mutual inductance depends on the antennae properties and the distance between two antennae. The intensity of the coupling decreases with increasing distance and results in reduced signal strength of the transmitted data. Consequently, the analysis of the signal strength allows concluding the distance between a tag and a reader. However, if the tag antenna is tilted away from the central axis by the angle φ, the induced voltage is smaller and follows the relationship given in equation $u_{o\varphi} = u_o \cdot COS(\varphi)$ (u_0 is the voltage that is induced when the coil is perpendicular to the magnetic field and uoφ is the actual induced voltage).

In order to apply proximity sensing to high frequency RFID systems information about the tag's orientation in relation to the reader antenna is required as well as information about the dimensions of the tag and reader antennae and the number of windings thereof. While the reader antenna and the tags used in our application remain unaltered, tilt angle and signal strength are subject to change. The orientation of a tag is derived from a tilt sensor or gyroscope that is directly attached to the RFID tag. The sensor data is transmitted along with the tag's ID. Based on the sensor data, the reader computes the deflection with respect to its own orientation. In combination with the signal strength of the received data, the reader estimates the distance of the tag from the reader antenna.

System Design and Implementation

Our RFID system comprises a tag, a reader, and a data processing unit (e.g. a notebook or a cell phone) that is attached to the reader. The reader and the reader antenna are commercially available. However, we built our own passive RFID tag with discrete components to provide an interface that allows attaching a tilt sensor. This is required because current passive tags do not provide any interfaces to attach additional devices. Our tag consists of a coil to pick up the reader signal, an analog front-end to demodulate and modulate the RFID signals, and a logic unit to interpret the demodulated data and to encode and send tag data. Additionally, our tag contains a tilt sensor to provide information about the deflection of the tag away from the parallel plane of the reader antenna as well as an external power supply, e.g. a battery. The external power supply is required because a tag built with discrete components draws considerably more energy than a regular tag. Our tag's power requirements exceed the energy that can be provided by the reader.

Analog Front-end to De-/Modulate RFID Signals

Our tag implements the widely used ICODE1 communication protocol developed by Philips. The ICODE1 protocol operates at 13.56 MHz. It is a readertalks-first protocol that uses amplitude modulation and pulse position coding on the reader-to-tag communication channel. The protocol uses a modulation index of 10% and, in standard mode, data is transmitted according the '1 out of 256' pulse position scheme. The value of a transmitted byte is encoded in the position of the pulse that is set to one of 256 possible consecutive positions. The data transmission is preceded by a start pulse of 9.44μs that signals the demodulator of the RFID tag that a new sequence of data is sent. The demodulator on the tag determines the bit position by measuring the elapsed time between the start pulse and the detection of the next pulse. Each bit has a length of 18.88μs. Dividing the measured time by the sequence of one bit results in the exact bit position and allows decoding the value of the byte. 4.833ms are required to complete the transmission of a single byte (excluding the start bit). Hence, the transmission of an entire RFID reader command, which consists of 8 consecutive bytes, takes 38.7ms.

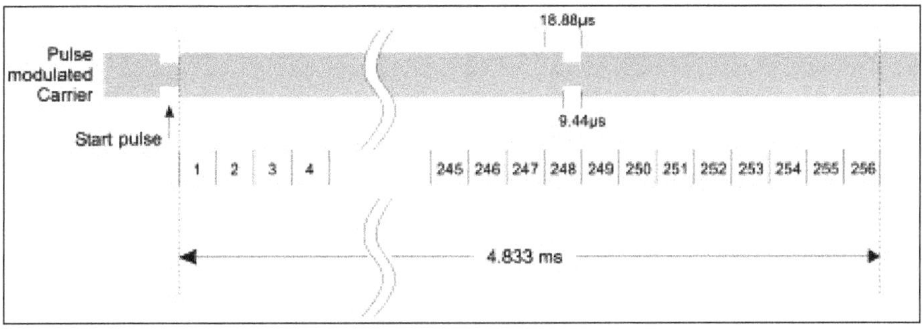

The value of a transmitted byte is encoded in the 256 possible consecutive positions for a pulse.

The analog front-end on the tag comprises an envelope detector to identify relative changes in amplitude. These changes indicate modulated bits. However, amplitude variations can also be caused by changes in coupling between a reader and a tag coil (e.g. tilting of the tag or a change in the distance to the reader). In order to separate amplitude variations that are caused by modulated bits during data transmission from those that are caused by changes in coupling, the output of the envelope detector is compared to a relative reference voltage. The reference voltage reflects variations in distance and orientation but its time constant is chosen so that the voltage remains unaffected by short pulses used for data transmission. Both, the envelope signal and the reference voltage are extracted by two RC filter elements. The RC elements are designed to meet the specifications of the ICODE1 air interface protocol with a 10% amplitude modulation and a signal drop of 9.44μs for a modulated signal. In addition, these RC elements account for negative peak clipping that is caused by the half-wave rectification.

The envelope signal and the reference voltage are compared by an operational amplifier element that generates a binary output based on the detection of a modulated signal. The output goes low as long as the signal is modulated and remains high otherwise. The system makes use of load modulation as described above to transmit data from the tag to the reader. A CMOS switch, controlled by a microcontroller, short-circuits the antenna coil according to a signal pattern that represents the encoded data.

Logic Unit and Signal Processing

A PIC16F88 microcontroller with an internal clock speed of 5MHz forms the core of the tag. The clock is not obtained from the carrier signal of the RFID reader but from an external oscillator. We made use of an external oscillator to avoid building a clock extractor thereby simplifying the design.

The PIC16F88 microcontroller offers two different methods to listen for changes on the comparator's output – polling and interrupts. While polling proved inadequate because it does not meet the timing constraints for the detection of two consecutive pulses, the interrupt-based approach performs well. The asynchronous interrupt is triggered by a falling edge at the comparator's output. When an interrupt occurs, the interrupt dispatcher reads the value of the timer that was started after the detection of the start bit. This allows determining the value of the byte transmitted by the reader. After the detection of 8 consecutive bytes, the main routine in the microcontroller code starts decoding the complete command.

The most restrictive time constraints occur when a pulse is sent in the first slot. This pulse follows 18.88μs after the falling edge of the start pulse. The detection of each pulse triggers an interrupt that is used to measure the elapsed time between consecutive pulses. However, the processing of an interrupt must be completed before the next pulse can be detected. This sets an upper time limit for interrupt processing of 18.88μs. Our optimized interrupt dispatcher requires 77 clock cycles, which corresponds to 15.4μs. Hence, our design fulfills even the most restrictive time constraints.

In response to a "selected read" or "unselected read" reader command, the tag responds with the data stored at the location specified in the reader command. When data is sent from the tag to the reader, the shortcircuiting of the tag's antenna generates only small variations of the voltage at the reader antenna. Therefore, a sub-carrier of 423.75kHz is used in accordance with the ICODE1 specifications to create modulation sidebands at the reader's antenna to simplify detection. Based on this sub-carrier, the tag data is Manchester encoded (idem). The transmission of a single bit requires 37.76μs and an entire byte requires 302.08μs. The data is sent in a random time slot while the length of the time slot depends on the number of data packets to be sent. The transmission of one data packet requires 1208.32μs.

Tilt Sensor

We use a dual-axis ADXL202E accelerometer to measure gravity. The accelerometer allows determining the tilt angle in one plane. For reasons of simplicity, our design is limited to the detection of deflections in the y-z-plane, but by attaching an additional tilt sensor or by using a gyroscope, the deflection in all three dimensions could be captured.

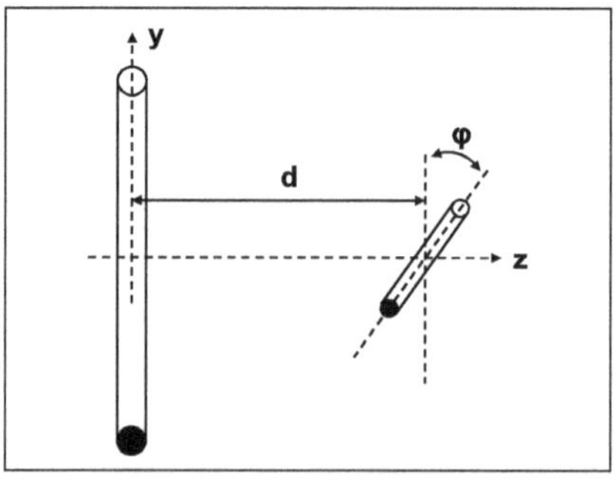

Cross-section through the reader and tag antenna, which are aligned on the coil axis. The tag antenna is tilted by the angle φ.

The analog outputs of the tilt sensor are connected to the A/D converter of the PIC microcontroller, which processes and transmits the data. The sensor values show a non-linear behavior of tilt angle and output value. On average, a sensor value unit corresponds to 1.0° to 1.8°. The standard deviation for a fixed tilt sensor is 2.8 units, or 3° to 5°. Due to the properties of the cosine function, small deflections away from the parallel plane do not result in a significant change of the actual inducted voltage.

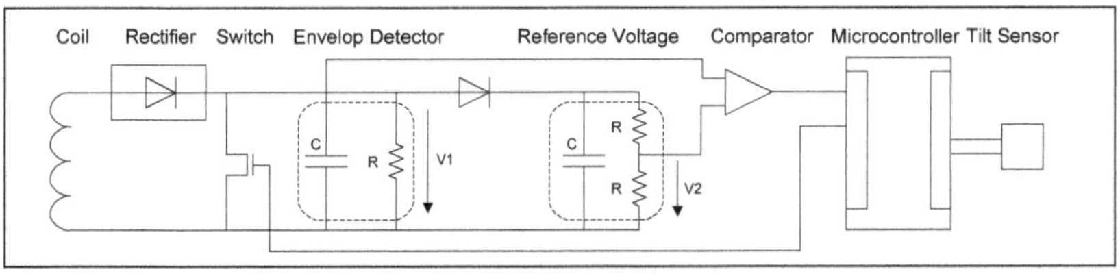

Schematic of the tag with the analog front-end, logic unit, tilt sensor, and CMOS switch for data transmission.

Distance Measurement

The measurement of a tag's distance from the reader antenna requires a three-step procedure. First, we analyze the signal strength of the data received from the tag. The signal strength directly relates to the distance between the tag and the reader antenna. Our RFID reader, a RIDEL5000 from Softronica, provides an interface to capture the signal strength of the received data. The signal strength of each data bit is sampled 8 times with a resolution of 12 bits. The amplitude of the signal strength is extracted with a simple peak extraction algorithm. In order to cope with signal strength fluctuations, we collect a set of 80 data values at once during a measurement period of about 60s to compute the average amplitude of the signal strength. In addition, we make use of a threshold-based noise filter to improve the data quality.

Second, the tilt angle φ that was transmitted to the RFID reader over the RFID communication channel is extracted from the received data. We compute the reduction in induced voltage caused by the tilt.

Third, we retrieve the absolute distance by comparing the transformed signal strength to a previously established look-up table. The look-up table was created with a tag in parallel to the reader and a steadily increasing distance with increments of 1cm. The signal strengths were recorded and stored in a lookup table along with their associated distances.

ULTRA HIGH FREQUENCY RFID

The Ultra High Frequency (UHF) band is the frequency range from 300 MHz to 1 GHz. RFID systems that operate in this frequency band are called Ultra High Frequency (UHF) RFID Systems. UHF RFID tags/readers usually operate at 433 MHz and from 860 to 960 MHz. UHF RFID tags operate using the far-field radiative coupling principle or backscatter coupling principle.

UHF RFID tags have a much higher read range when compared to LF and HF tags. The high data transfer rate makes UHF RFID tags ideal for applications that need to read multiple items at once, such as boxes of goods as they pass through a door into a warehouse or racers as they cross a finish line.

The read range of UHF RFID tags can be as high as 50 feet which is much higher than that of LF and HF tags. The longer read range opens up a set of new applications for RFID such as electronic toll collection and parking access control.

UHF tags do have some downsides, one being that they operate at a higher frequency and thus have shorter wavelengths which attenuate fairly quickly and are also very sensitive to interference. These tags cannot easily be read while attached to objects containing water and animal tissues because water absorbs UHF waves. The UHF tags also do not work well with metals, as they get detuned when they are attached to metal objects.

Table: UHF RFID Tag Properties.

Tag Type	Passive / Battery Assisted / Active
Frequency	433 MHz and 860 to 960 MHz
Read Range	Up to 50 feet
Read Speed	Very Fast
Read Multiple Tags Simultaneously	Yes
Works with Metal	No
Works with Water	No
Cost	Very Low

UHF RFID systems function within these designated bands due to communication protocol standards put in place by GS1 as well as frequency regulatory standards set by individual countries/regions. These standards and regulations are an integral part of RFID. If they didn't exist:

- Devices could interfere with each other and would be either inoperable or have very poor efficiency as they compete for bandwidth.
- Widespread adoption of the technology would be challenging because systems would be operating in disparate bands and have differing protocol/signaling standards.

UHF RFID systems do not communicate through magnetic coupling; rather, they use passive backscatter modulation according to GS1 EPCglobal UHF Class 1 Gen 2 specifications (which are ratified as an ISO-18000-6C standard).

UHF waves are, on average, around 33 centimeters (or one foot) in length; but, in the higher frequencies, like 2.45 GHz, they can be as small as 12 centimeters in length. UHF RFID is typically deployed in one of two types of RFID systems – passive or active.

Passive UHF RFID

The defining quality of passive UHF RFID systems (when compared to active UHF RFID systems) is the way in which the RFID tags function. Passive UHF RFID tags rely on passive backscatter modulation to function and have no additional power source. In short, this means that the RFID reader sends its energy through the antenna as RF waves to the UHF RFID tags in order for them to become energized and respond back to the reader. The response is called backscatter because the tag scatters back a portion of the energy that it receives from the reader. Because there is no additional power available to the tag other than what is provided by the reader, these systems have a maximum read range of around 30 meters.

UHF RFID passive tags have some of the lowest costs due to widespread adoption. Volume quantities of UHF RFID tags can dip as low as $0.10 cents a tag while maintaining an average read range of 2-5 meters. UHF RFID passive tags have decreased in cost substantially over the years allowing companies and individuals to use low-cost, label tags as disposable asset identifiers.

Passive UHF tags are very sensitive to both liquids and metal. This is because the UHF waves reflect off, refract within, or absorb into non-RF-friendly materials. Though performance degradation in the presence of liquids and metals may be difficult to overcome, interference can be mitigated by analyzing the application environment and employing techniques to overcome obstacles, such as installing RF shielding and using metal-mount RFID tags. Passive UHF tags come in many form factors and are usually subdivided into categories like rugged, high-temperature, label tags, high memory, etc. The tags tend to last the lifespan of the application unless damaged by wear and tear, and because they do not have batteries, their lifespan is not dependent on a power source.

The hardware for UHF RFID system requires a more significant investment than HF and LF system equipment. Readers range between $450 for a low-cost reader up to over $2,500 for a rugged handheld RFID reader.

Passive UHF RFID Applications

Passive UHF RFID systems are used in hundreds of different applications such as tool tracking, IT asset tracking, race timing, and laundry management. New applications for these tags are being discovered frequently due primarily to the tags' long read range and low cost.

Active UHF RFID

Active UHF RFID systems do not rely on passive backscatter to function; instead, they use an internal battery as a power source. Because these tags contain a battery, they do not need to be energized by an RFID reader; instead, they proactively beacon at predetermined intervals. These beacons announce the presence of the tag to any readers that are within range of the tag. Active RFID readers detect the beacons from the tags and pass them on to a host system for processing.

Active UHF RFID tags can have a very long read range due to the internal battery, with beacons that can be detected by readers over 100 meters away. The signal boosting power received from the internal battery also helps these tags to overcome any materials that usually hinder RF waves like metal and water. Certain active RFID tags are engineered to withstand harsh environments like extremely low temperatures and rugged applications, including being able to continue beaconing while being buried in snow or dirt.

Active RFID tags are relatively expensive because they contain an internal battery and other electronics components. Depending on their features (e.g. Wi-Fi, Bluetooth, GPS), these tags cost $20 per tag and up. One drawback of active tags is their limited lifespan. Because they rely on a battery for power, these tags (like their batteries) can only last about 3-5 years (depending on beacon rate). While that timeframe works well for some applications, other applications may require the tags to be replaced.

Active UHF RFID Applications

Active UHF RFID applications are frequently used in industries like oil and gas, transportation, and vehicle tracking. Because active tags beacon, they are easier to read while moving and are ideal for tracking cargo containers and vehicles. Additionally, tracking items like pipes and construction equipment is one of the more popular uses for active RFID because laydown yards encompass very large areas. Active tags can be installed on materials and large assets outdoors and read by handheld or vehicle mounted readers in order to obtain their location information.

CHIPLESS RFID

Radiofrequency identification (RFID) is a wireless technology for identification and tracking of objects, animals, persons, etc. which has moderately penetrated the market, as compared to optical identification systems (based on barcodes or QR-codes). Despite the fact that RFID offers significant advantages over optical systems, such as larger reading distances without the need of line-of-sight, high data storage capacity, and the possibility to read multiple tags simultaneously, the cost of traditional RFID tags (i.e., tags equipped with a silicon integrated circuit or chip), is excessively high for many applications involving low-cost items. This applies even to passive tags, fed by the interrogation signal, where batteries are not present. Thus, the typical cost of passive UHF-RFID tags is situated in the vicinity of several eurocents, much beyond the cost of optical barcodes or QR codes. Since the cost of chipped RFID tags is mainly dictated by the presence of the chip, the implementation of planar encoders, able to replace the chip in the tag, has been an object of intensive research in the last years, giving rise to the so-called chipless-RFID technology. Thus, in chipless-RFID tags, the tag is equipped with a planar encoder (typically a printed pattern containing the ID code) and sometimes with an antenna for communication with the reader. By using low-cost conductive inks, the price of the tag can be substantially reduced. However, with chipless-RFID it is not possible to be as competitive as with chipped-RFID in terms of system performance (read range, data capacity, etc.). Indeed, the main research activity relative to chipless-RFID technology in the last years has been oriented towards increasing the data storage capacity and density of the tags.

Chipless-RFID Tags based on Time Domain

Time Domain Reflectometry based Tags (TDR)

The former chipless-RFID tags were inspired by the operation of RFID tags which encode information in the time domain. Such tags were implemented on surface acoustic wave (SAW) technology and basically consist of an electro-acoustic transducer connected to an antenna. The working principle is shown in figure. An electromagnetic pulse is sent by the reader which is received by the antenna and transformed into acoustic energy by the transducer. Then, the acoustic wave propagates through the substrate, reflecting part of it, in those positions where there is a reflector. Such reflections are converted back into electromagnetic energy and transmitted to the reader. This working principle is generally designated as time domain reflectometry (TDR), and the ID code is contained in the echoes generated by the reflectors. It is important to mention that chipless-RFID tags based on SAW technology provide high information capacity (up to 256 bits). Nevertheless, such tags are not competitive in terms of cost, since tag price may vary around 10–20 euro cents.

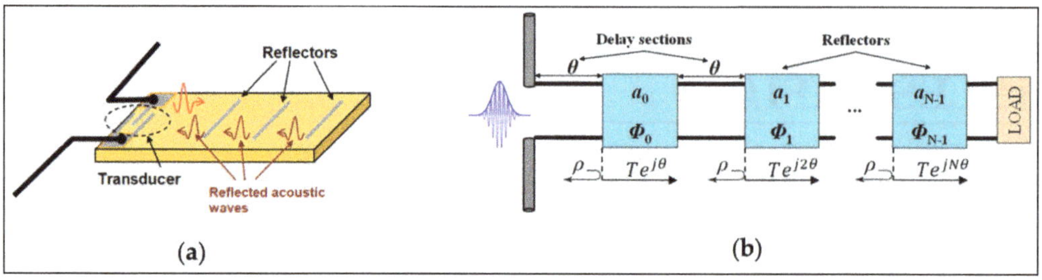

Working principle of a chipless radiofrequency identification (RFID) system based on (a) surface acoustic wave (SAW) technology and (b) delay lines. In both cases, the functionality is known as time domain reflectometry (TDR).

One solution, with a very similar operating principle, consists of implementing chipless-RFID tags composed of a delay line (transmission line), on conventional microwave substrates, which contain discontinuities (as reflectors) or complex impedances at certain positions. Basically, a pulse is sent by the reader which propagates through the delay line. Such a pulse is reflected at those points where there is a discontinuity. In this way, the identification code is constituted by the reflections or echoes produced in the tag. It is important to mention that the speed of sound is 340 m/s, one million times smaller than the speed of light (c). This aspect implies either implementing transmission lines by means of large lengths to produce measurable delays or very narrow pulses in order to avoid overlapping between reflected pulses. Such requirements are difficult to achieve in practice; therefore, the bit encoding capability is generally limited in chipless-RFID tags based on time domain.

On-off Keying (OOK)

On-off keying (OOK) encoding is considered to be the simplest encoding method in time domain-based chipless-RFID systems. Basically, the logic states '0' and '1' are determined by the absence or presence, respectively, of a reflection of the interrogation signal in a specific time interval.

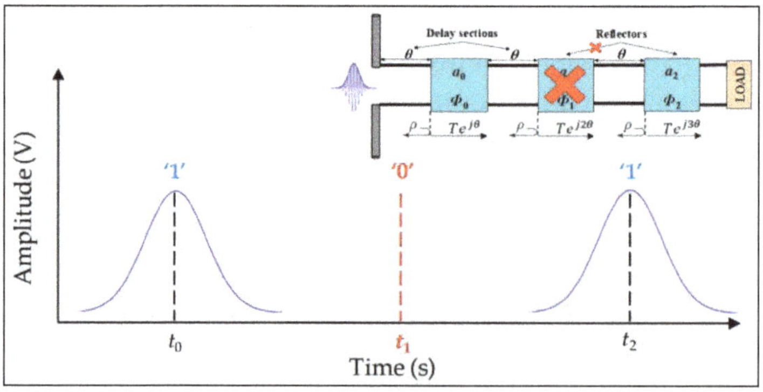

Working principle of on-off keying encoding.

The OOK encoding method was reported for the first time. The tag is composed of a meandered coplanar waveguide transmission line (CPW) with surface mount device (SMD) capacitors inserted in parallel at predefined positions. Such capacitors or discontinuities act as reflectors. A pulse of short duration, specifically 2 ns, which propagates through the transmission line CPW, is sent by the reader. The presence of a capacitive discontinuity at certain positions produces a mismatching that results in a reflected pulse which is transmitted back to the reader. The presence of such reflection is interpreted as the logic state '1', whereas its absence is associated to logic state '0'. An example of a measured 4-bit tag.

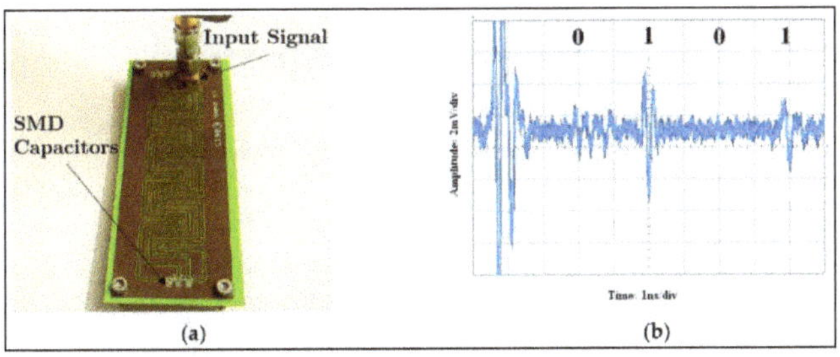

(a) Chipless-RFID tag proposed by Zhang et al. (code '0101'); (b) measured code '0101'.

Zheng et al. and Botao et al, respectively, replaced the SMD capacitors by patches, in order to obtain a fully planar structure. In this case, such patches were printed during tag fabrication and disconnected from the microstrip transmission line. Consequently, the identification code of the tag is programmed by means of the connection of certain patches to the delay line. In order to reduce the size of the tag by approximately a factor of four, using the same substrate the pulse duration of the interrogation signal was reduced to 0.4 ns. An example of a measured 8-bit tag.

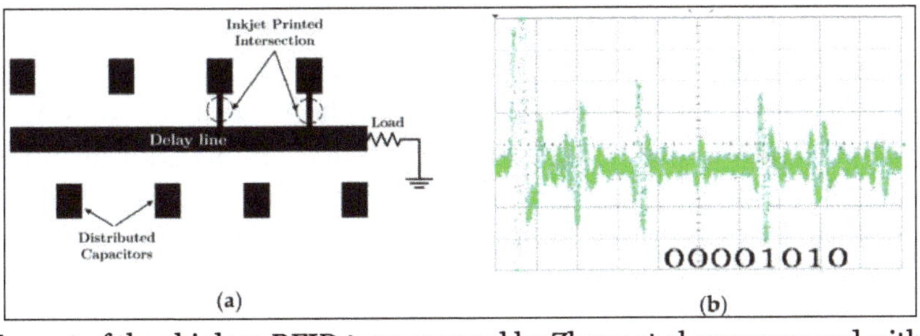

(a) Layout of the chipless-RFID tag proposed by Zheng et al. programmed with the code '00001010' and (b) measurement of the code '00001010'.

In a different strategy to implement chipless-RFID tags based on time domain with OOK coding was reported. The main novelty is the fact that the delay line is used in transmission, rather than in reflection. The input signal is divided into two branches, a short and straight line (without delay), and the other one a longer and meandered line (with delay). The two branches are connected by a circulator (which acts as an isolator) to deflect the second pulse at the appropriate time. Thus, the temporal position of the pulse is modified with the position of the isolator, and consequently, the encoding of the tag. In order to clarify the encoding strategy, figure show the principle of code generation, based on the superposition of delayed signals, and a simulated code example, respectively.

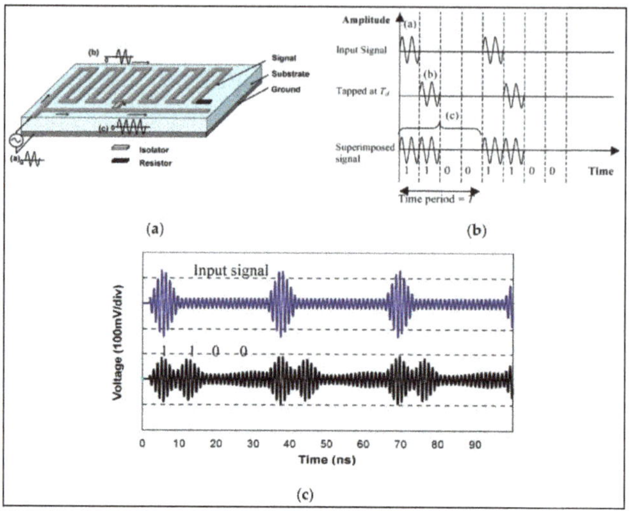

(a) Sketch of the delay line proposed by Chamarti et al.; (b) binary code generation by the superposition of delayed lines; (c) simulated input and output signals.

On the other hand, in 2012, a novel strategy based on delay lines able to support magneto-inductive waves (MIW), was proposed by Herraiz et al. MIWs are slow waves that propagate through a chain of magnetically coupled resonators. Such slow waves exhibit a group velocity (v_g) around c/100 (where c is the speed of light in vacuum), hence maximizing the group delay (τ_g). Thus, for the same delay, shorter delay lines can be achieved (as compared to conventional lines). For the implementation of the delay lines, split ring resonators (SRRs) with rectangular topology were used to maximize the coupling. A microstrip transmission line coupled inductively to the first resonator of the chain of SRRs, acts as a transducer between the microstrip mode and the MIW mode of the delay line. Finally, the additional microstrip transmission lines are used as reflectors, and coupled to specific SRRs of the chain in order to generate a specific identification code. Thereby, the position of the reflectors is different for each tag, giving rise to different temporal responses, as appreciated in figure c. The main limitation of this strategy is the fact that the high losses of the MIW delay lines prevent the implementation of tags with high information capacity.

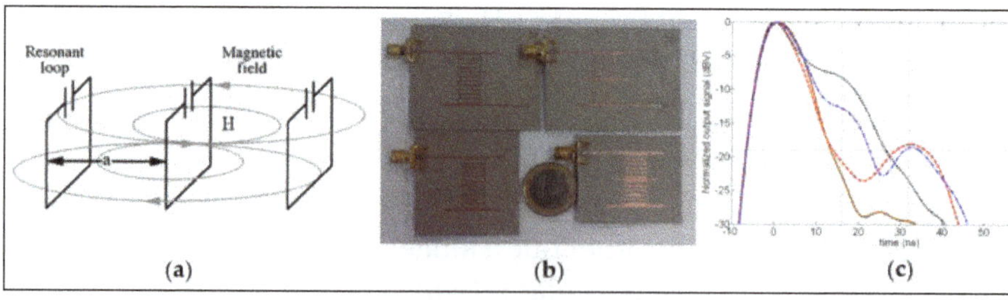

(a) Magneto inductive waveguide which consists of a set of capacitively loaded loops, magnetically coupled to each other. (b) photograph of the fabricated set of 2-bit chipless RFID tags and (c) measured response.

Pulse Position Modulation

Pulse position modulation (PPM) encoding consists of the variation of the position of a pulse within a certain temporal window. Specifically, PPM encodes n bits with a single pulse, which occupies one of the 2n slots of a temporal window. The advantage of this encoding, with respect to OOK, is that it requires a smaller number of reflectors. Nevertheless, a longer time span is necessary in order to achieve the same information capacity.

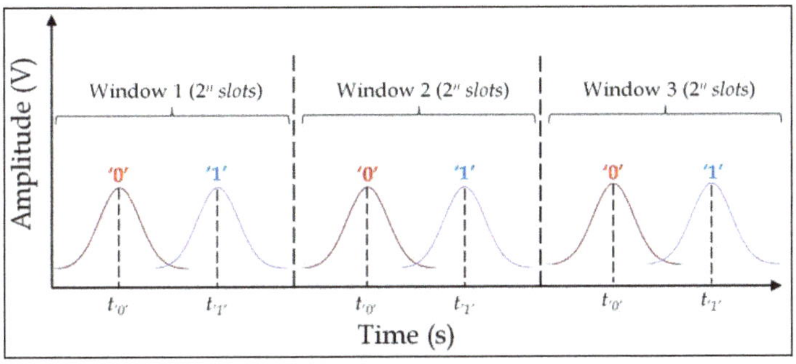

Working principle of pulse position modulation encoding.

Most of the works reported in the literature that use PPM encoding use dispersive structures which exhibit a frequency-dependent group velocity. Thus, a broadband signal propagating through such structures experiences a different group delay for each spectral component. This structure can be implemented by means of C-section all-pass networks, which are composed of two coupled lines, whose electrical length is $\lambda_g/4$ at the frequency where the group delay is maximum. Therefore, the frequency position of such maximum depends on the length of the structure, as depicted in figure b. Consequently, in the time domain, for the corresponding frequency, a pulse delay equal to the group delay was implemented.

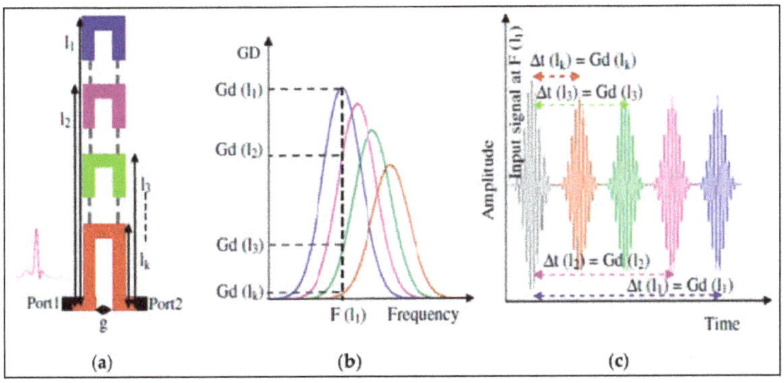

Principle of encoding by means of group delay: (a) structure of the chipless-RFID tag; (b) group delay vs. frequency for different lengths of C-sections, and (c) the corresponding time domain response.

In 2010, a 3-bit tag based on cascaded C-section dispersive structures was implemented for the first time by Gupta et al. Such a tag was implemented in order to provide maximums in the group delay at 3, 4, and 5 GHz. In this case, each pulse can provide two different positions within a temporal window, depending on its encoding ('0' or '1').

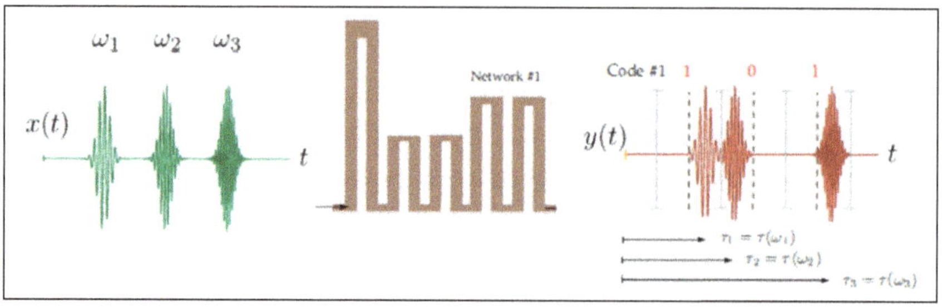

Chipless-RFID tag proposed by Gupta et al. fed with three Gaussian pulses with frequency ω_1, ω_2, and ω_3 and an example of PPM encoding.

Phase Modulation

In order to increase the information density of the tag, a different coding method based on the phase modulation of the input pulse to the tag was proposed by Mandel et al. Moreover, such tags include the use of composite right/left handed (CRLH) artificial delay lines, also used in. The use of these lines is justified by their lower group velocity, as compared to conventional delay lines. This aspect allows for the implementation of shorter delay lines for a specific delay. In this case, complex loads of different values are used for the purpose of modifying the phase of the reflected pulses. As a result, with four different elements it is possible to achieve four different phase values (2 bits for each reflector). The number of codes that can be generated is, therefore, 4^k (where k is the number of reflectors of the delay line).

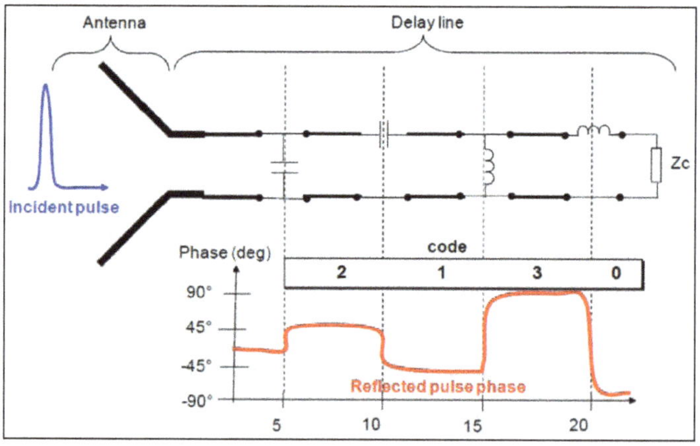

Temporal encoding based on the phase modulation of the input pulse to the tag.

Time Division Multiplexing based Tags

Time division multiplexing is a novel and unconventional strategy to implement chipless-RFID systems based on time domain, which is of particular interest in secure paper applications. Nevertheless, the working principle is completely different than those of the aforementioned chipless-RFID systems. Such a strategy was reported for the first time by Herrojo et al. in 2017. With this approach, the number of achievable bits is only limited by the area occupied by the tags and the ID code is contained in the amplitude modulated signal generated by the tags. Such tags typically consist of a chain of either resonant elements or metallic strips, etched or printed at predefined and equidistant positions on a dielectric substrate. Basically, in a tag reading operation, the tag is mechanically displaced over the sensitive part of the reader.

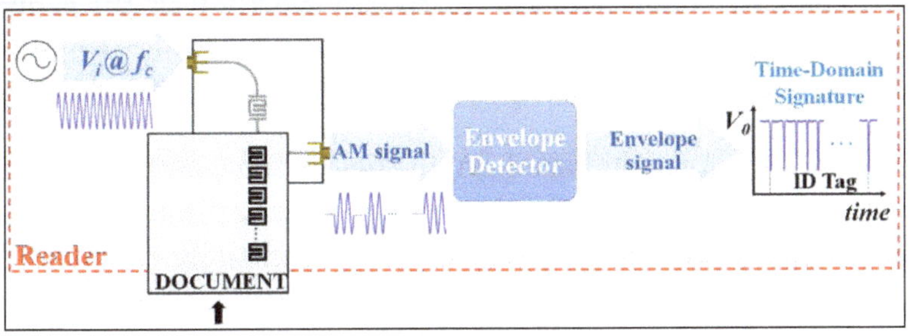

Sketch of the working principle of time domain chipless-RFID systems based on near-field coupling and sequential bit reading.

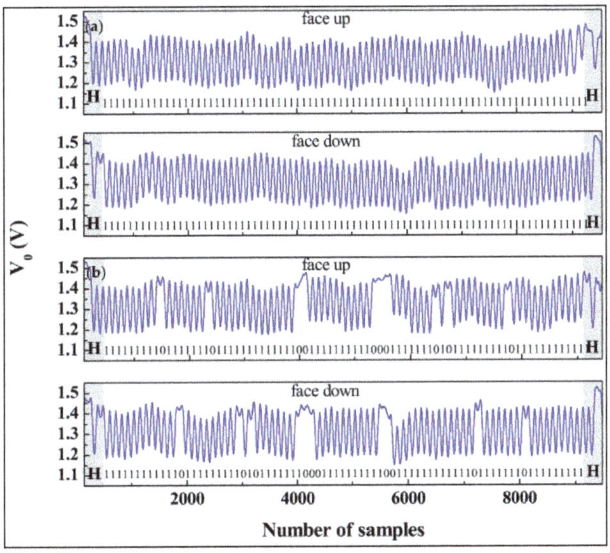

(a) Measured envelope of the inkjet-printed 80-bit tag with all bits set to the logic state '1' and (b) 80-bit programmed tag with the indicated code.

The reader is a planar microwave structure, typically a resonant element, coupled to a transmission line and fed by a harmonic signal. By positioning the resonator of the sensitive part of the reader at a short distance from the tag, the tag motion modulates the amplitude of the feeding signal at the output port of the transmission line (a sketch of the working principle of the proposed chipless-RFID system is depicted). This amplitude modulation is due to the electromagnetic coupling between the resonator of the reader and the resonators or metallic strips of the tag, which modulate the magnitude of the transmission coefficient at the frequency of the feeding signal. With respect to encoding, it can be done either by cutting those elements of the tag associated to the logic state '0' or by its presence/absence at predefined equidistant positions. The former strategy is important in order to reduce costs, since it is possible to fabricate thousands of identical tags with a single mask and programming such tags in a later stage. In, such a strategy was demonstrated since the authors implemented an 80-bit tag by means of inkjet printing on a conventional paper with all bits set to '1' logic state, and then programmed with different codes by cutting those resonators associated to the logic state '0'. Such an approach provides, in terms of performance, high capacity and relatively small size at the expense of near-field tag reading. Nevertheless, the required proximity between the reader and the tag is not an issue in certain applications, such as secure paper (of interest to avoid counterfeiting or copying, e.g., of official documents, etc.).

Recently, an alternative approach based on all-dielectric electromagnetic encoders was proposed, avoiding the use of metallic elements (either resonant element or strips). In this case, the AM signal is generated by means of permittivity contrast. Specifically, in, the functionality of the system was demonstrated by implementing a chain of 10 apertures on the dielectric substrate RO4003C with permittivity and loss tangent of 3.55 and 0.0021, respectively. The sketch of the working principle is depicted in figure whereas the normalized measured envelope functions of three different codes are shown in figure.

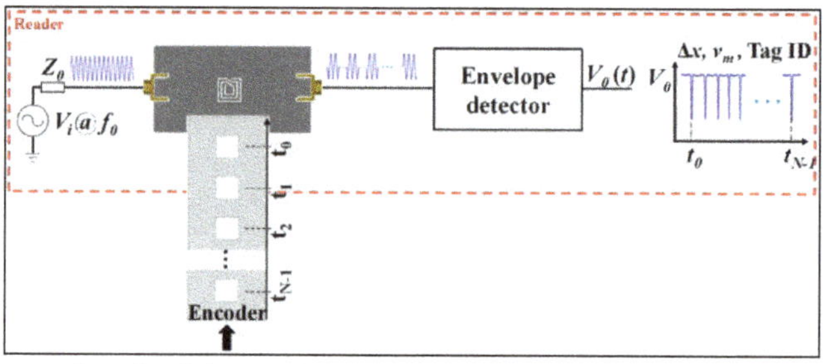

Sketch of the working principle of the all-dielectric electromagnetic encoders based on permittivity contrast.

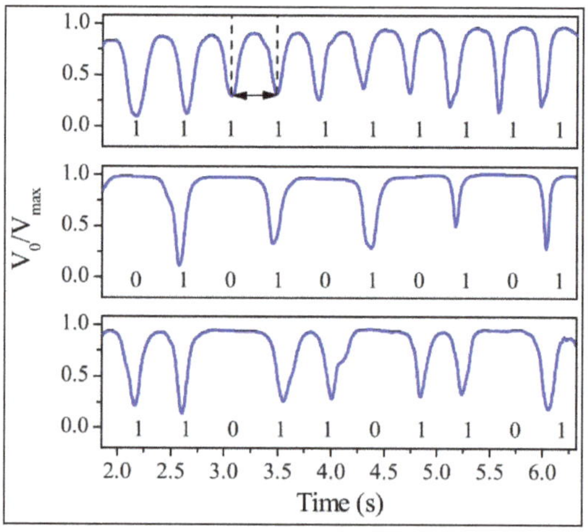

Normalized measured envelope functions of the indicated codes.

Chipless-RFID Tags based on Frequency Domain

Chipless-RFID tags based on frequency domain, also known as spectral signature barcodes, are implemented with resonant elements tuned at different, and predefined, frequencies. Such frequencies are distributed within a certain frequency band, which must be covered by the interrogation signal. Typically, each resonant element provides a bit of information, and the encoding is determined by the absence or presence of singularities in the amplitude and phase of the frequency response of the tag. According to the type of interaction between the interrogation signal and the tag, frequency domain-based tags can be classified as retransmission-based and backscattered-based tags.

Retransmission-based Tags

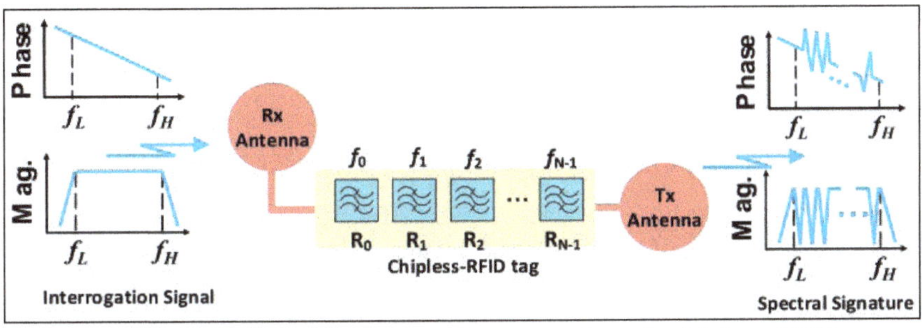

The working principle of retransmission frequency domain chipless-RFID systems.

The working principle of retransmission-based tags is depicted in figure above. Such tags were designed, implemented, and patented by Preradovic et al. for the first time. Typically, these chipless-RFID tags are equipped with a transmitting and receiving antenna, which are cross- polarized in order to communicate wirelessly with the reader. Such antennas are used for the reception of the interrogation signal and the transmission of the spectral signature of the tag. A tag to be highlighted is the one implemented by Preradovic et al.

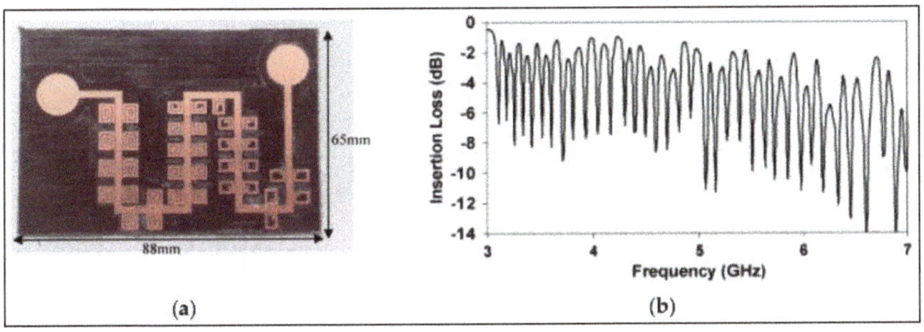

(a) Photograph and (b) measured response of a 35-bit tag.

As shown in figure below the tag was implemented by a meander microstrip transmission line (to reduce the size) loaded with multiple resonators, specifically 35 spiral resonators, hence encoding 35 bits in a bandwidth and a surface of 4 GHz and 88 × 65 mm² (including antennas), respectively. With regard to the frequency response of the tag, it presents 35 transmission zeros in the magnitude of the spectral signature, introduced by each of the elements which are part of the tag.

Backscattered-based Tags

In backscattered chipless-RFID based tags, the resonant elements provide the spectral signature through singularities in the response of the radar cross section (RCS).

The main advantage of this strategy is the fact that the use of antennas is not necessary, hence the size of the tag is reduced. The working principle is illustrated in figure.

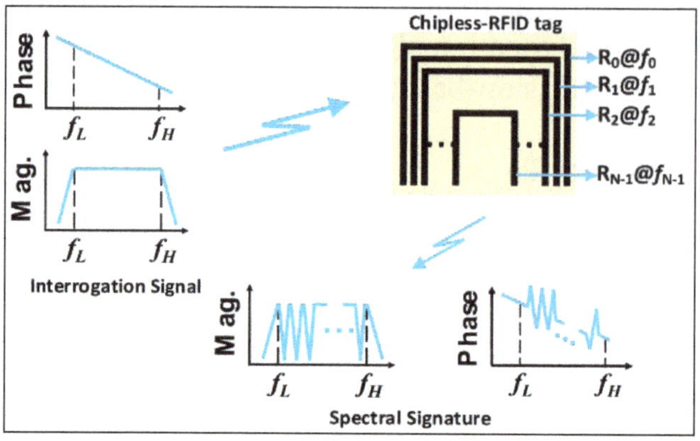

The working principle of backscattered frequency domain chipless-RFID systems.

In 2005, Jalaly et al. proposed a very simple solution consisting of multiple dipoles with a variable capacity implemented on a substrate. The main purpose of this strategy is to create multiple resonances in the bandwidth of the tag. Similar strategy, implemented by the same authors, consists of using dipoles with variable lengths, making the structure completely planar. The tags are implemented in the frequency range from 2.4 to 5.8 GHz and the information capacity is 5 bits.

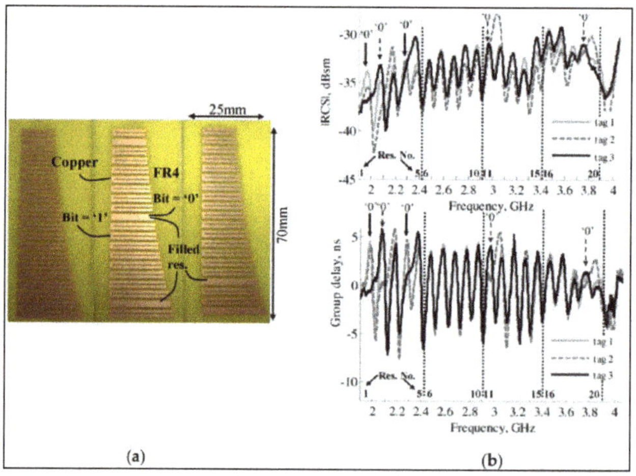

(a) Photograph and (b) measured radar cross section (RCS), and group delay of three different 20-bit tags.

A tag to be highlighted is the one shown in figure above and implemented by Vena et al. Such a tag is composed of 20 resonant C-shaped elements with resonance frequencies between 2 and 4 GHz, and a size of 25 × 70 mm². A peak appears in both the magnitude of the RCS and in the group delay at the resonance frequency of each resonator. With regard to the encoding, the logic state '0' is achieved by shifting the resonance frequency of the resonators (C-shaped resonator) to higher frequencies. Basically, such a frequency shift is achieved by short-circuiting the resonators. The fact that it is not necessary to remove them from the tag is an important aspect, since it opens the possibility of manufacturing all the tags with all the resonators short-circuited, programming them, and removing the short-circuit with a laser, for example.

The main limitation of encoding information in the frequency domain is the bandwidth required to accommodate a large number of bits. Moreover, increasing the information capacity of the tags implies increasing the number of resonant elements, and, consequently, their size. In recent years, in order to minimize such limitations, different research groups have focused their efforts on designing tags able to encode information in more than one domain simultaneously, for example, frequency-phase, frequency-amplitude, etc. Such systems are known as hybrid systems, and the information capacity of the corresponding tags can be substantially increased, as compared to one of the frequency domain tags of this section.

Hybrid Tags

The main objective of hybrid systems is to assign more than two logic states to a single resonant element. In most cases, the approach to achieve such multi-state behavior is by encoding the information in the frequency. Thus, the resonance frequency of each resonator can vary between different values within a predefined frequency window. Obviously, with this strategy, the number of states is limited by the bandwidth assigned to each resonant element. To clarify this concept, figure shows a chipless-RFID tag based on three circular rings. Such a tag operates in the frequency band from 3.1 to 10.6 GHz (i.e., in a bandwidth of 7.5 GHz which has been divided into 30 MHz slots). Therefore, there are 250 slots to be shared between three rings (80 for each one), and, consequently, 80^3 = 512,000 different codes (19 bits) can be generated in an area of 9 cm².

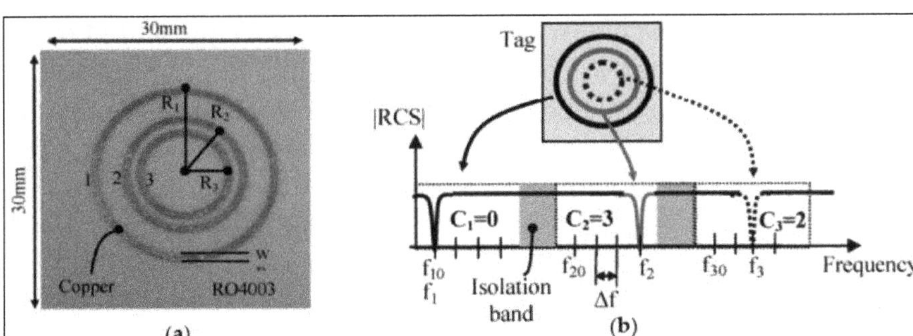

(a) Photograph of a chipless-RFID tag constituted by three circular rings and (b) the coding principle.

Another approach to achieve a multi-state functionality is to encode the information in the amplitude level of the transmission zeros introduced by the resonant elements of the tags or in the level of the magnitude of the resonance peaks of the RCS. An example of such encoding is shown in figure below. The information was encoded in the magnitude level of the resonance peak introduced by the C-shaped resonator. In this case, each resonator provides 2 bits (four states) of information. The bandwidth and the area of the tag are around 3 GHz and 7.2 cm², respectively.

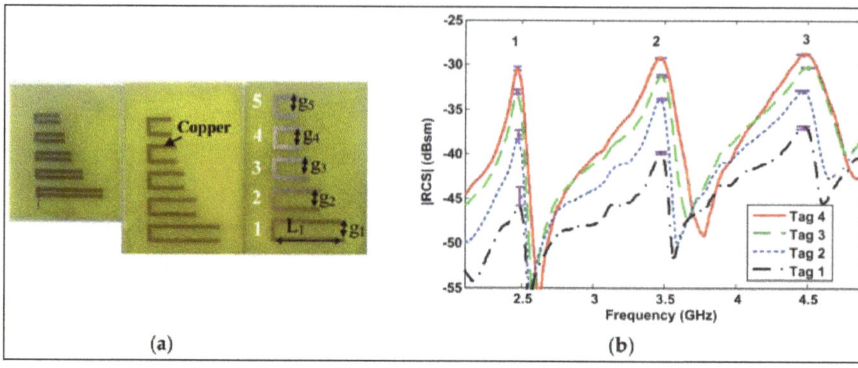

(a) Photograph of three chipless-RFID tags proposed by Vena et al. and (b) the measured frequency response.

In 2011, Vena et al. presented a tag which encodes the information in two independent parameters, the phase deviation and the position of the frequency within a given frequency window.

Such a tag is constituted by five C-shaped resonators printed on a FR4 substrate with resonance frequencies within the frequency band from 2.5 GHz to 7.5 GHz. Each resonator exhibits a peak and a notch controlled, basically, by the resonator length and g/L ratio, respectively. In the example of figure b, the logic state '00' is assigned to the response with a narrower phase deviation and a peak resonance at 2.5 GHz while the logical state '01' corresponds to the response with a wider phase deviation and maintains the same peak resonance. The logic states '10' and '11' correspond to a narrow and wide phase deviation, respectively, but in this case to the peak resonance at 3 GHz. It is clear that, with this example, 2 bits of information can be assigned to each resonator. However, varying the dimensions of the resonant elements, it is possible to increase the information capacity of each particle. In fact, in the work presented by the authors, each resonator provides 24 states. Therefore, it is possible to generate 24^5 = 7,962,654 different codes (equivalent to 22.9 bits) in a size of 8 cm^2 with only five resonators.

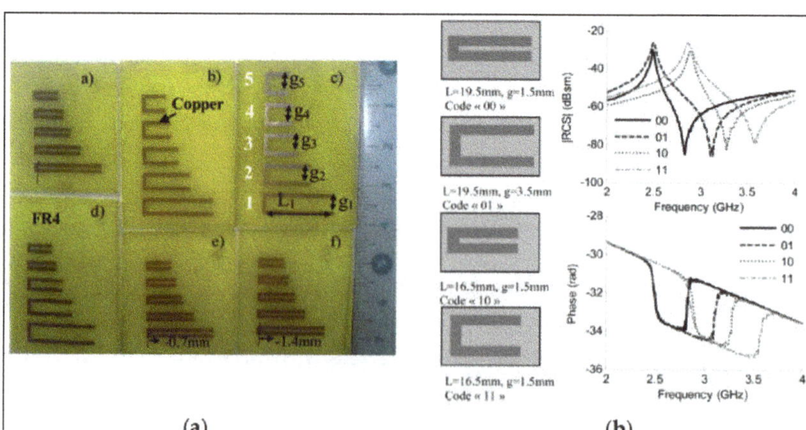

(a) Photograph of six chipless-RFID tags proposed by Vena et al. and (b) the measured frequency response.

Another possibility, proposed by El-Awamry et al. in, consists of a hybrid coding which combines the frequency position of a notch and its bandwidth. This coding technique consists of three different bandwidths (BW_1, BW_2, and BW_3) and three different frequency positions (f_r, $f_r + \Delta f_r$, $f_r - \Delta f$), where f_r is the resonant frequency of the resonant particle and Δf is the frequency offset used to encode. In order to achieve three different bandwidths, the tag uses different coding elements: a dipole, a square ring, and a patch. Consequently, the control of the position in frequency of the notch is done by varying the dimensions of the resonant elements. With this strategy, 48 bits are achieved (4 bits per resonator) in a size of 20 cm² and using a bandwidth of 3 GHz.

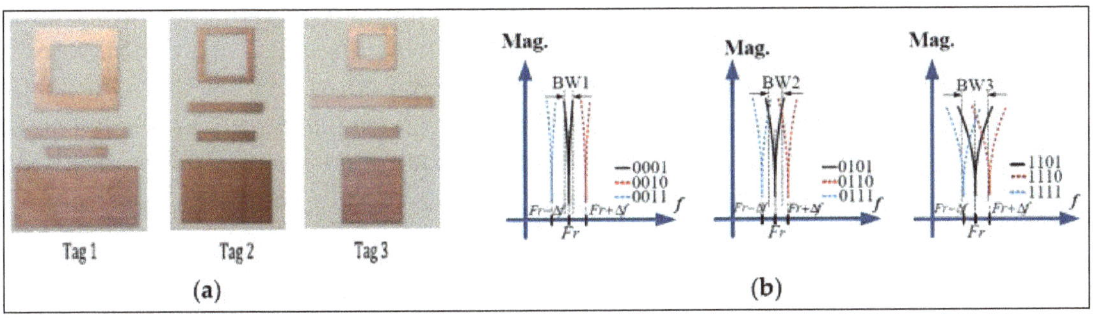

(a) Photograph of three chipless-RFID tags proposed by El-Awamry et al. and (b) the working principle of the proposed encoding.

Polarization diversity is also useful to increase the information encoded in a given surface and bandwidth. This method is used in order to multiply by a factor of two the information capacity of the tags in the same spectral bandwidth. Such tags consist of two sets of rectangular slot resonators etched on rectangular metallic patches with different polarizations (vertical and horizontal), and in turn, each set of resonators is divided in two patches. It is important to mention that the slots etched in each patch are conveniently placed in order to reduce the coupling between resonant elements. On the other hand, in the tag reading operation, the tag must be illuminated with a plane wave orthogonally polarized. Therefore, the reader consists of two dual-polarized antennas, one to transmit the interrogation signal and the other one to receive the spectral signature of the tag. With this strategy, as a proof-of-concept demonstrator, the authors encoded 16 bits within an area and bandwidth of 3.06 cm² and 5 GHz, respectively.

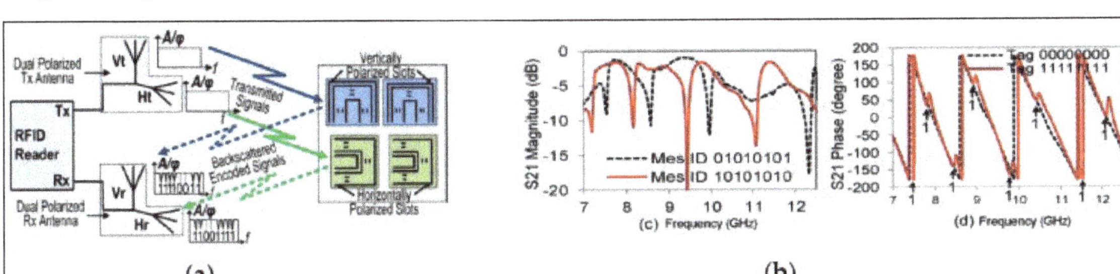

(a) Chipless-RFID based on polarization diversity proposed by Islam et al. and (b) measured frequency response of a tag with the indicated code.

RADIO FREQUENCY IDENTIFICATION AND SENSING TECHNIQUES

Radio Frequency Identification (RFID) sensors, integrating the features of Wireless Information and Power Transfer (WIPT), object identification and energy efficient sensing capabilities, have been considered a new paradigm of sensing and communication for the futuristic information systems. RFID sensor tags featuring contactless sensing, wireless information transfer, wireless powered, light weight, non-line-of-sight transmission, flexible and pasteable are a critical enabling technology for future Internet-of-Things (IoT) applications, such as manufacturing, logistics, healthcare, agriculture and food. They have attracted numerous research efforts due to their innovative potential in the various application fields. However, there has been a gap between the in-lab investigations and the practical IoT application scenarios, which has motivated this survey of this research to identify the promising enabling techniques and the underlying challenges.

Radio Frequency Identification (RFID) technology, as a key enabling technique of the Internet of Things (IoT) sensing layer now finds applications in a wide spectrum of fields for data integration and management, including human identification, logistics and retail, access control, parking management, indoor localization, etc. Compared to the other alternatives, such as barcodes and QR codes, the radio frequency-powered identification approach featuring contactless, wireless powered, non-line-of-sight, read- and writeable, light weight, and multiple tag simultaneous reading allows the connected "things" to be identifiable for further data communication and integration. The functions of RFID are to collect RF energy from the interrogator with its antenna, either Ultra-high Frequency (UHF) or High-Frequency (HF), activate the RFID chip in the tag, and transmit an ID code back to the interrogator, where the ID code is a fixed number used as a unique identifier of a "thing". The features of the "things" corresponding to their identities can be saved into a database and updated in real-time for data management to meet the needs of different applications.

However, since the RFID chip is an Integrated Circuit (IC) powered by RF energy, components with sensing capability can also be potentially integrated into RFID tags for simultaneous identification and sensing purposes. The integration of RFID tags with sensing components could eventually deliver identification and sensing capability in a wireless powered, contactless, and non-line-of-sight way.

Simultaneous Wireless Information and Power Transfer (SWIPT), different from the conventional wired- or battery-powered sensing, has become a new paradigm of sensing and communication, which could potentially reshape the future rich-sensing IoT world. On the one hand, due to the wide coverage and mobility of RFID interrogators, the measurement of RFID sensor tagged "things" is no longer limited to specific locations. On the other hand, since RFID sensors can be fully passive, there is no need to frequently change the batteries of the RFID sensors like in traditional wireless sensor nodes. Therefore, the information sensing procedure of RFID sensors becomes much more flexible and convenient, and its applications can be extended into a wider field.

Firstly, they have been widely applied in industry for the sensing of temperature and humidity, strain, pressure, steel corrosion and cracks, concrete structure, pipeline integrity monitoring, etc. Secondly, they are also popular in healthcare, in devices such as wearable and implanted sensing devices for glucose monitoring, blood pressure, intraocular pressure, and on-skin monitoring discrimination of breath anomalies etc. The above examples are some exemplary cases of investigations, and the research and applications are not limited to them. The investigations of RFID sensing techniques can be mainly divided into four categories: (1) radio frequency energy harvesting efficiency; (2) the integration of RFID and sensing techniques; (3) chip-less RFID techniques; (4) RFID sensor network technology.

Although significant technical progress regarding the research and application of RFID sensing techniques has already been achieved, there is still a gap between in-lab investigations and practical applications. In addition, there are also technical issues in the integration of RFID technology with sensing components that have not been sufficiently addressed.

The Fundamentals of RF Energy Harvesting and RFID Sensor Techniques

The source power of energy harvesting can be in different forms, such as mechanical movement and vibrations, solar energy, thermal energy, electromagnetic power, etc. Since the early 1990s, when Tesla and Hertz proposed the concept of wireless energy harvesting, research on wireless energy collection has become more and more extensive in the scientific community. According to the fundamentals of RFID, radio frequency energy harvesting can be divided into inductive coupling and backscattering. The HF RFID working at the carrier frequency 13.56 MHz transmits and receives power with near-field inductive coupling, and the UHF working at the carrier frequency 840–960 MHz deals with power transmission and reception with far-field backscattering. This investigation focuses on the HF and UHF RFID sensor techniques, for which, new investigations and novel technical solutions have been widely reported in the recent few years.

HF Inductive Coupling and HF RFID Sensor Techniques

HF RFID sensor devices transmit or receive both energy and data through inductive coupling, which is achieved by the alternating magnetic field between the coil antennas of the interrogator and the tag. The alternating magnetic field is generated by the inductor-capacitor (LC) resonant tank, the structure and the equivalent circuits of which are as depicted in figure.

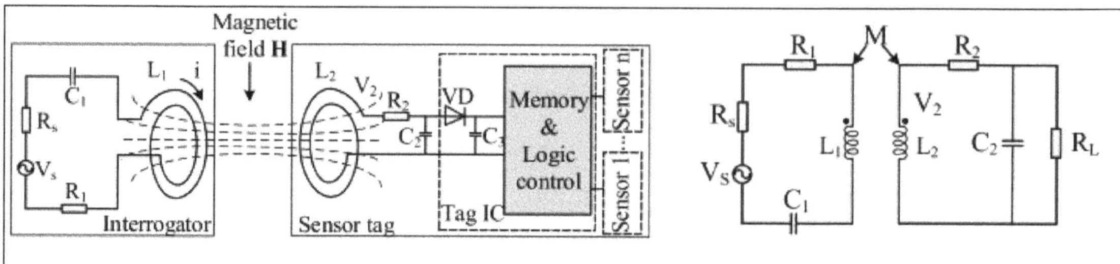

(a) Inductive Coupling of HF RFID Interrogator and Tags (b) Equivalent Circuit of Inductive Coupling

As shown in figure, V_s is the RF source in the interrogator, L_1 and C_1 constitute a resonant tank (resonance to the frequency V_s) for power transmission, and the resonant frequency is f_o, normally 13.56 MHz for HF RFID. The tag antenna L_2 and C_2 constitute the receiver, which works at the same frequency. The resonant frequency can be calculated as follows:

$$f_o = \frac{1}{2\pi\sqrt{L_1C_1}} = \frac{1}{2\pi\sqrt{L_2C_2}}$$

The inductive voltage v_2 on the tag antenna is directly proportional to the change rate of magnetic flux Ψ through the coils, which can be represented by:

$$\upsilon_2 = -\frac{d\psi}{dt} = -N_2\frac{d\phi}{dt} = -N_2\frac{d}{dt}(\int B \cdot dS) =$$

$$-N_2\frac{d}{dt}\left[\int \frac{\mu_0 i_1 a^2 N_1}{2(a^2+r^2)^{3/2}} \cos\alpha dS\right]$$

where N_2 and N_1 are the number of windings of the tag and interrogator antenna coils, Φ is the magnetic flux of each winding coil, B and S are the strength of magnetic induction and the area of the coils, i_1 is the current of the interrogator coil, a is the radius of interrogator coil, r is the distance between the two coils, and α is the angle between B and S.

$$\text{In, } f_o = \frac{1}{2\pi\sqrt{L_1C_1}} = \frac{1}{2\pi\sqrt{L_2C_2}}$$

when the angle between B and S is 0^0, υ_2 can be calculated by the following expression:

$$\upsilon_2 = -\left[\frac{\mu_0 N_1 N_2 a^2 S}{2(a^2+r^2)^{3/2}}\right]\frac{di}{dt} = --M\frac{di}{dt}$$

where M is the mutual induction coefficient, which can be calculated with:

$$M = \frac{\mu_0 N_1 N_2 a^2 S}{2(a^2+r^2)^{3/2}}$$

Formulas above provide a clear relationship between the power received and the key parameters of the two antennas and their relationship. The voltage V_2 can be processed with rectifier and regulator circuits to power the RFID chip and sensor module for data acquisition and information transmission.

UHF Backscattering and UHF Sensor Techniques

Different from a HF RFID system, the operating mode of a UHF RFID sensor device is radar backscattering, consisting of the UHF interrogator, RF transmission path, RFID tag and sensor module. The performance of the power transfer is determined by the parameters of the main components, including the radiation power and gain of the transmission antenna and receiving antenna, the wavelength of carrier signal, and the distance between the transmission and receiving antennas.

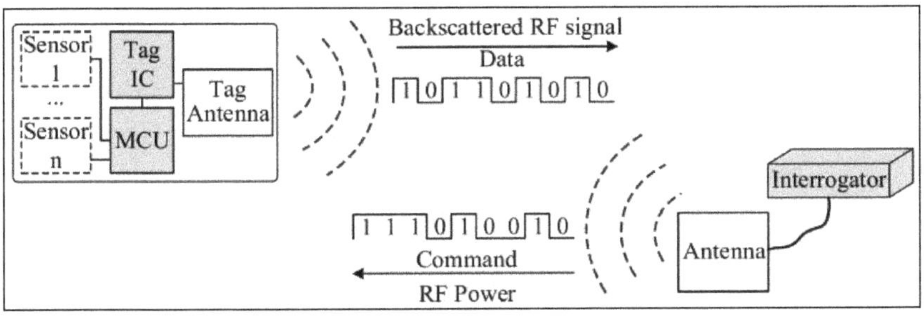

The Fundamentals of UHF RFID Sensors.

Interrogator-to-tag RF Power Transfer

As a passive device, the RFID sensor tag receives the interrogator's power to complete the sensing and data transmission procedures. According to the mathematical relationship for RF transmission given by Friis equation, the power density of RF energy at a distance R from the antenna of interrogator S can be represented by:

$$S = \frac{P_T G_T}{4\pi R^2}$$

here P_T and G_T are the power and gain of the transmission antenna. Suppose the effective dimension of a tag's receiving antenna is A_E defined as $A_E = \lambda^2 G_{Tag} / (4\pi)$, the gain of the receiving antenna is G_{Tag}, and the wavelength of RF carrier signal is λ, then, the received power P_{Tag} can be expressed by:

$$P_{Tag} = S \cdot A_E = P_T \left(\frac{\lambda}{4\pi R} \right)^2 G_T G_{Tag}$$

UHF RFID Power Reflection

The RF power reflected by the tag is directly proportional to the tag's Radar Cross-Section

(RCS) σ. The power reflected by RFID tag P_{Back} can be denoted with equation below, where *S* is the power density, and the power density S_{Back} can be denoted by equation below, P_T and G_T are the transmission power and gain of tag antenna. Suppose G_R is the gain of interrogator antenna, the effective dimension of the receiving antenna is $A_w=\lambda^2 G_R/(4\pi)$, and the power received by the interrogator P_R can be represented by equation below:

$$P_{Back} = S\sigma = \frac{P_T G_T}{4\pi R^2}\sigma$$

$$S_{Back} = \frac{P_T G_T}{(4\pi R)^2 R^4}$$

$$P_R = S_{Back} A_w = \frac{P_T G_T G_R \lambda^2 \sigma}{(4\pi)^3 R^4}$$

The tag's information, including the tag ID and sensor information, is transmitted to the interrogator by modulating the reflected carrier RF waves. Evidently, the gain of the antennas, the distance between antennas, and the tag's RCS are key parameters determining the reading distance and efficiency of the resulting RFID sensor measurements.

Chip-less RFID Sensors

Chip-less RFID sensors simply change the radar profile of the RFID tags to transmit sensor data without the need for complex integrated circuits, the structure of which is as shown in figure . When the receiving antenna of the tag receives the signal transmitted by the interrogator, the resonant circuit selects signals of different frequencies to generate a specific resonant frequency and a different resonant amplitude, and then returns the data to the interrogator through the transmitting antenna of the tag. The interrogator then parses the amplitude and phase changes of its specific frequency signal. Finally, the signal ID is encoded according to different spectrum characteristics after data processing. When the chip-less tag is placed on a Material Under Test (MUT), the resonant frequency will drift as the sensor data changes, as shown in figure.

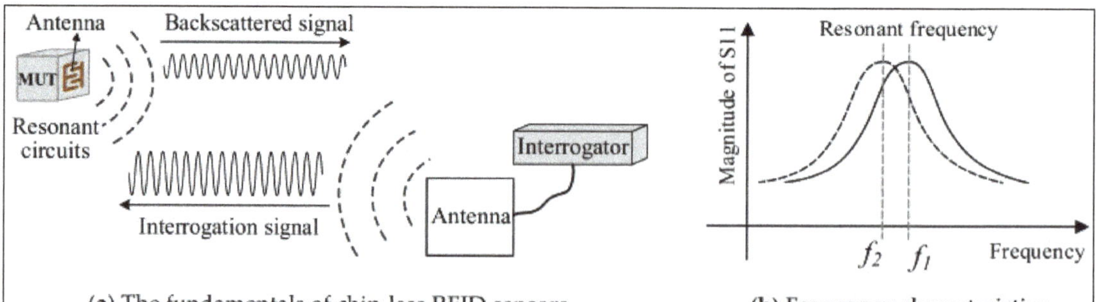

(a) The fundamentals of chip-less RFID sensors (b) Frequency characteristics

Chip-less RFID Sensors—Fundamentals and Frequency Characteristics.

The function of chip-less RFID is determined by the antenna which is normally designed for some particular applications. The ID of chip-less RFIDs are defined by the characteristic frequencies of the tag at a series of specific values. Since some physical features of the materials under test can be reflected by the resonant circuits in the form of capacitance and inductance. The data storage capacity of the chip-less RFIDs is far less than the chipped ones. However, due to silicon chips, the cost of chip-less RFIDs is significantly lower than the chipped ones. Therefore, chip-less RFID has become a competitive choice for low-cost sensing and identification applications.

In addition to HF, UHF and chip-less RFID sensors, some applications utilize commercial tags for object distance measurement and localization, which may find use in applications like robot localization and navigation, product localization in large warehouse and logistics, etc.

RFID Sensor Techniques—Technical Progresses and Innovative Applications

Compared to other environmental energy harvesting approaches, radio frequency is a relatively efficient approach which can obtain more available power in a low-energy environment. In addition, RF energy harvesting integrated with information transmission, termed as Wireless Power Communication (WPC) and SWIPT, has enabled many innovative applications by the wireless powering, identification and communication. Especially in recent years, with the continuous progress of IoT application techniques and the increase of low-power sensing and micro-controller devices, the integration of sensor techniques with RFID has attracted much research interest and made remarkable progress. This section aims to provide a timely summary of the recent technical progress and innovative applications.

Technical Progress of RFID Integrated Sensing

Novel Antenna Designs: The Front End of RFID Sensing Techniques

The RF energy harvesting circuit is mainly composed of an antenna, a rectifier, a voltage multiplier and an energy storage device as shown in figure, in which the antenna captures the RF energy in the electromagnetic field, then rectifier converts RF energy into a DC voltage through a rectifier, and finally stores the increased voltage of the voltage multiplier in the energy storage device.

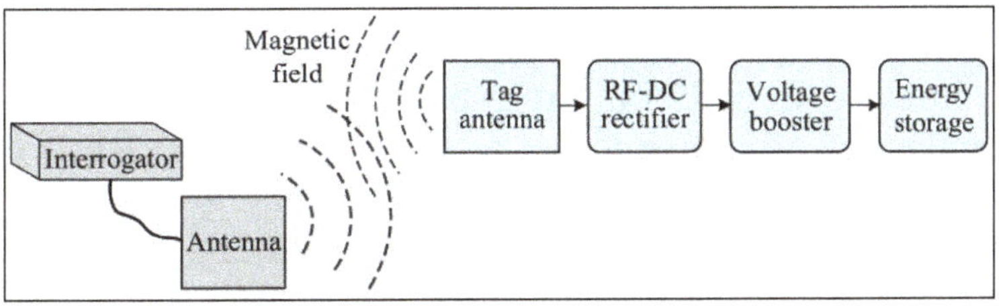

The Diagram of a RF Energy Harvesting Circuit.

The commonly used RFID antennas are shown in figure. The antenna in figure is normally applied to the HF systems, and the antennas in figure–d are usually applied to UHF systems. High gain antennas can improve the conversion efficiency and acquire more power, so high gain antennas are preferred. The performance of antennas can be evaluated with parameters including gain, frequency band, radiation pattern, polarization, physical size or fields of application.

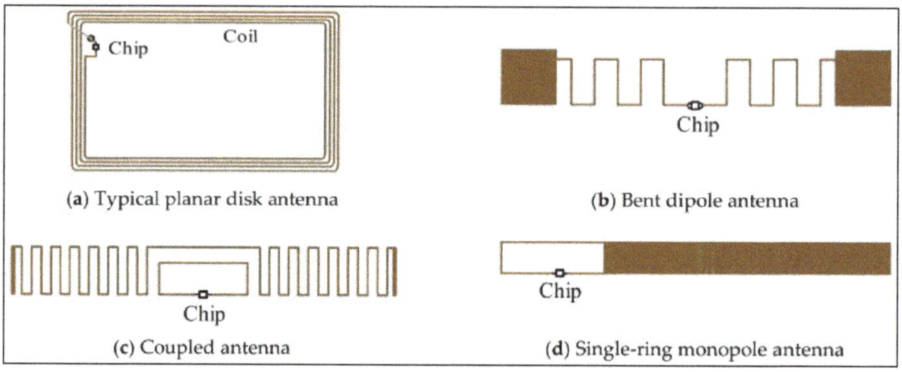

Commonly Used RFID Antennas.

Located in the front of a RFID sensing device, the antenna is a key component determining the performance of a RFID sensor, including its reading distance, operation speed and the size of sensor module. Therefore, novel antenna design to enhance the performance of sensor devices and for different application scenarios becomes a hotspot. The main contributions focus on miniaturization of the size, foldable antennas, 3D antennas, metallic surface mountable antennas, multi-band antennas, and omnidirectional and directional antennas, etc.

With respect to metallic surface mountable antennas, Chen proposed a miniature UHF RFID antenna design for metallic object applications, consisting of two rectangular patches electrically connected through via to the ground plane and a non-connected inter-conductive layer to form an RFID tag antenna. The miniature design is achieved by inserting a conductive layer into the antenna structure to increase the capacitive reactance of the antenna. The overall size is 32 × 18 × 3.2 (mm) and the maximum reading range of RFID tag placed on a metallic object is about 1.5 m. 3D antennas have also become an interest in order to pursue omnidirectional patterns in orientation-independent RFID applications, and many new designs are proposed. Kruesi et al. introduced a miniaturized folded meander line 3D cubic antenna for wireless sensor network (WSN) and RFID application in environmental sensing with a 3D dimension of 3 × 3 × 3 cm, which may find applications in smart packaging by integrating it into a cube's hollow interior. In addition, multi-band antennas, and omnidirectional and directional antennas are attractive topics as well for different RFID applications. By increasing the working bands of antennas to collect energy over multiple frequency bands, the output voltage of the RF energy harvester can be increased. Keyrouz et al. designed a three-frequency antenna (900 MHz, 1800 MHz, and 2.45 GHz) to collect

RF energy, and the energy collection efficiency increased significantly. However, due to the increase in the number of antennas, the dimension of circuit area was enlarged. Liu et al. proposed a capacitively loaded, inductively coupled fed loop antenna with an omnidirectional radiation pattern for UHF RFID tags, with a small size of 43 × 43mm, the measured maximum reading range of the prototype is 9.5 m with a total transmitted power of 4.0 W effective isotropic radiated power (EIRP). The omnidirectional radiation pattern on the horizontal plane makes the proposed tag insensitive to be mounted on different target objects.

Since RFID sensing devices are used for different purposes and application scenarios in different ambient environments, the design of antenna should be customized to the requirements of the particular applications.

However, for RFID sensors, the antenna design of a typical RFID tag is versatile and requires the impedance matching for the maximal reading range. But, due to the sensors and various devices, energy consumption has increased significantly. In order to obtain stable RFID sensor data, the RFID sensor antenna can also be configured as a multi-port to collect natural energy to enhance the energy harvesting of the self-powered wireless sensor, as shown in figure.

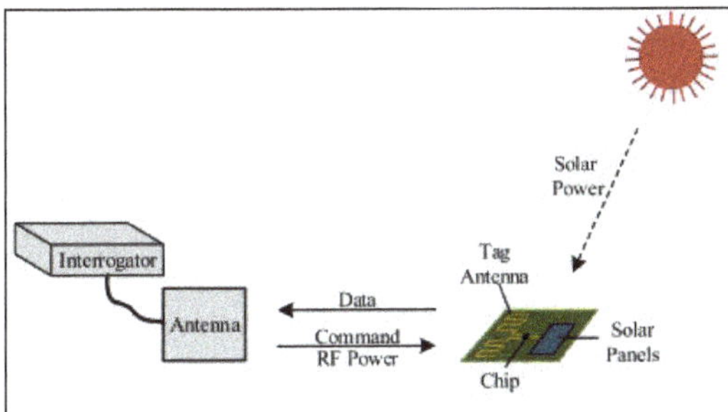

(a) Solar energy integrated RFID sensor

Different forms of natural energy
Solar Energy
Heat Energy
Motion
Friction, etc.

(b) Energy forms that can be integrated with RFID

RFID sensor with multi-port energy harvesting.

RF-to-DC Conversion and Power Management

Since the RFID chip and sensing module require DC power for data acquisition and transmission, the RF power needs to be converted to DC to provide the power supply. RF energy harvesting normally uses a multi-stage voltage rectifier to obtain high output voltage, and various rectifiers designed to perform the conversion and improve energy harvesting efficiency are presented in. Half-wave, full-wave, bridge and differential rectification methods, as shown in figure, are the most commonly used solutions. Bakhtiar et al. proposed a high sensitivity CMOS rectifier designed with standard threshold voltage devices. The rectifier is designed and laid out in a standard 0.13 μm CMOS process, allowing it to operate even when the input RF power is weak. Ye and Fan describe a high efficiency rectifier circuit for passive UHF RFID applications. The simulation and measurement show efficiency of 30.7% and 15% at low input power level respectively.

The rectifier was fabricated in a standard 0.18 μm CMOS process and its core circuit occupied 0.169 × 0.137 mm² silicon area. Ouda et al. proposed a cross-rectifier that can extend the dynamic range of RF power. The rectifier can achieve wireless power supply at different distances, which reduces the problem of reverse leakage without reducing its sensitivity. This design can be used in UHF RFID applications. On-chip measurements show that when the load is 100 kΩ, the sensitivity is −18 dBm at a voltage of 1 V, and the peak conversion efficiency of RF power to DC power is 65%. Yi et al. used a CMOS process in the design of a multi-stage rectifier, which uses a diode with a very low threshold voltage to connect the MOS transistor for energy conversion. However, since a MOS transistor is used, the threshold voltage is increased due to the bulk effect. Liu et al. used a novel diode-connected MOS transistor, and based on this, designed an N-level current rectification circuit, which has higher conversion efficiency and higher output voltage than a rectifier composed of a conventional diode and MOS transistor. However, the threshold voltage of the MOS transistor increases as the usage time increases. Rabén et al. developed a theoretical model for diode-connected MOS transistors with a threshold cancellation technique, and the derived design equations illustrate the tradeoff between the voltage drop and the reverse leakage of the diode. A rectifier was designed and implemented in a 0.35-μm CMOS process, and cadence simulation results of the PCE and the voltage conversion efficiency showed good agreement with the model. Besides UHF, there are also several RF-DC converters operating at 13.56 MHz. For instance, high-efficiency RF-DC converters exhibiting −4 dBm and 1.2 dBm sensitivity, respectively, are presented. A HF RF-DC rectifier that uses reverse loss reduction to ensure high output current, even if the DC-DC boost converter limits sensitivity. Colella et al. proposed a four-stage RF-DC converter for HF RFID that is fully integrated using 350 nm CMOS technology. Pelliconi's two-stage charge pump is used to increase the voltage. Experimental verification showed it can work normally with an RF input power of −19 dBm.

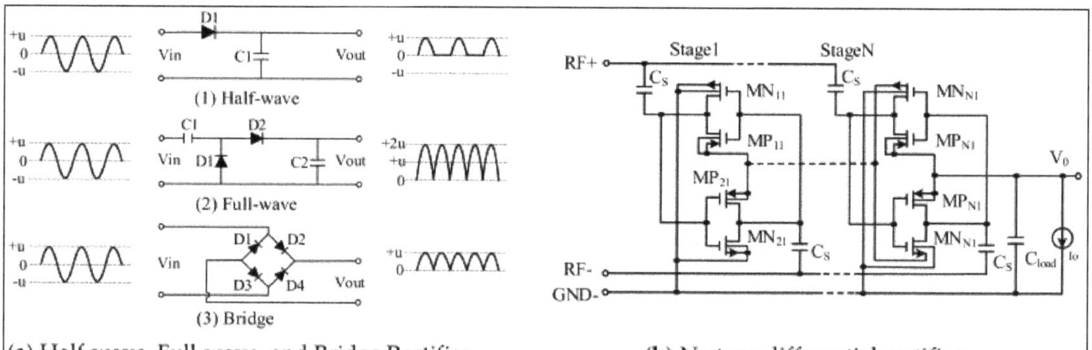

(a) Half-wave, Full-wave, and Bridge Rectifier (b) N-stage differential rectifier

Half-wave, Full-wave, and Bridge Rectifier.

Since the collected RF energy provides the power supply for the operation of the whole RFID sensing module, a voltage multiplier is expected to increase the voltage value and reduce the energy storage time. However, the output voltage of the voltage multiplier changes with process variations and temperature fluctuations (work time is too long or environmental impact). Yuan et al. studied the threshold voltage compensation scheme to improve the effects of voltage multiplier temperature and process variability. Chouhan et al. designed a new cascading method for voltage multiplication circuits, which allows the rectifier to have better power conversion efficiency at lower RF power.

Printable and Flexible RFID Techniques for Sensor Fabrication

Flexibility is an important advantage of RFID tags and RFID sensors, which allows the sensor tags to be able to pasted onto the object under test. The ink-jet printing process has been introduced to the fabrication of RFID sensors. Printed electronics is a new technology that combines traditional printing technology with electronic technology and applies traditional printing technology to electronic manufacturing. Compared with traditional electronic component manufacturing methods, printing technology is more stable in an oxidizing environment, and has the advantages of cost efficient, flexibility, and environmentally friendly. Among them, the contact type mainly includes flexographic printing, offset printing and screen gravure printing, and the non-contact type is mainly inkjet printing. The principle is as shown in figure.

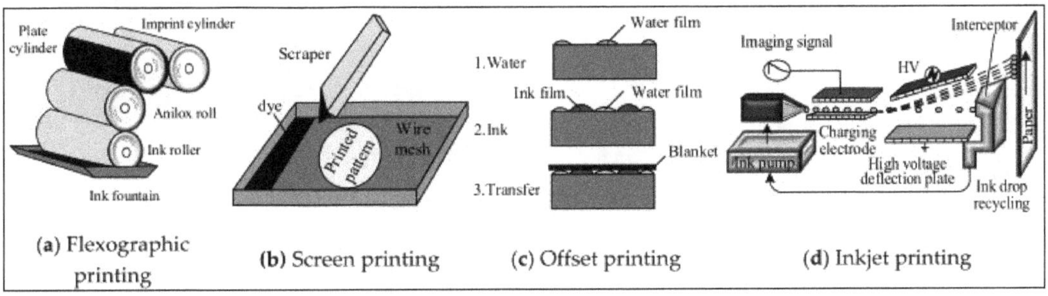

Four Typical Printing Techniques.

Flexible RFID printing is mainly used to fabricate chip-based RFID antennas and chip-free RFDs, which print nanoparticles such as graphene, silver, or copper on photographic papers, plastics or textiles. In the past ten years, the scarcity of resources, the rapid updating of products and the deteriorating ecological environment have made the market put forward low-cost, flexible and green requirements for modern technology. As an electronic substrate, paper substrates have the advantages of low cost and mass production, which has attracted wide attention. At the same time, although nanoparticle conductive inks are produced in a wide variety of materials (such as copper (Cu), gold (Au), graphene, carbon nanotubes (CNT)), silver nanoparticle ink is the main choice for the electronic conductive trace of printed paper.

Paper substrate inkjet-printed RFID sensors have been a prevalent way to fabricate flexible sensors. For example, Kim et al. presented a paper substrate silver nanoparticle-based printed sensor fabrication solution and provided two proof-of-concept sensors designs with evaluations:

- Chip-less CNT gas sensor.
- UHF dual-tag capacitive haptic sensors.

Sharif et al. presented a low cost, conductive ink printed small loop integrated with meandered dipole used as an inductive load, which was also connected with RFID chip for metallic can measurement in a smart refrigerator system. Borgese et al. introduced a chip-less RFID humidity sensor based on finite artificial impedance surface (AIS) inkjet printed on a thin sheet of commercial coated paper, composed of three concentric loops thus obtaining three deep and high-Q nulls in the electromagnetic response of the tag. Salmeron et al. reported two printed UHF RFID tags capable of measuring temperature and humidity, using the SL900A IC on-chip temperature sensor and capacitive humidity sensor fabricated on polyimide substrate:

- Inkjet-printed array of capacitive humidity sensors.
- Screen printing interdigitated capacitive sensors.

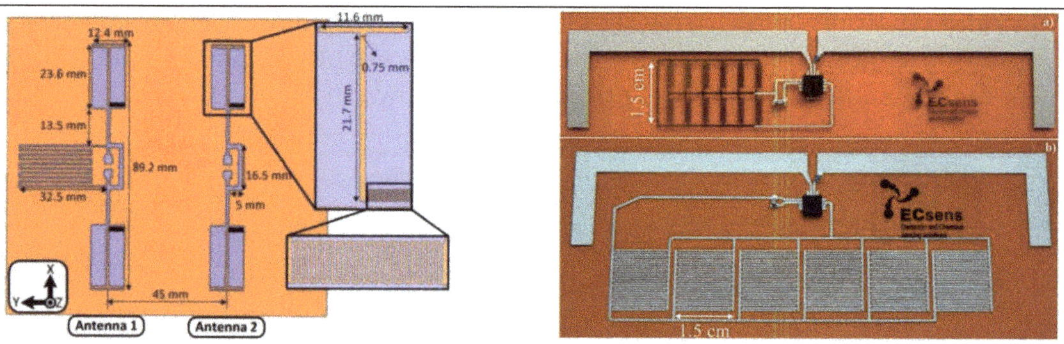

(a) Printed dual-tag capacitive haptic sensors: 1) sensor tag, and 2)

(b) Array of capacitive sensor by 1) inkjet-printing and 2)

Inkjet Printed RFID Sensors.

In the above applications, the printed antennas are connected to an RFID chip and sensing components to allow RF-powered sensing capabilities. The advantages of printed RFID electronics are:

- Sensors can be made flexible by choosing flexible substrates.
- The cost of RFID sensor fabrication can be effectively reduced.
- Sensing components can be designed as a printable structure.

These advantages allow the RFID sensors to be conveniently fabricated for different investigations and applications with reasonable low cost.

However, the size of the RFID sensor is still a limitation for many practical applications. Since the area of the RFID sensor chip is small, the size of the entire RFID sensor is determined by the size of the antenna. In order to study RFID sensors of small size, the use of textile technology to weave RFID sensors are introduced. Textile integrated RFID has the advantages of low cost, durability, and the separation of textile and RFID chip manufacturing, making mass production fast. Kalhnayer et al. studied the textile transponder system, in order to achieve a high read/write distance, the antenna uses a textile-based design, but the washability needs to be improved. Vieroth et al. used a flexible substrate on the coupling module package to obtain the flexibility of the package. The test results showed that the package with silver surface finish can withstand all test conditions, even washed at 60°C. Alonso-González et al. designed a three-layer woven structure that converts the antenna of an RFID tag into a woven type. The label can be processed directly into the garment, making it widely used in the apparel sector. The integration of RFID with textiles is an effective means to obtain the flexibility of RFID sensor devices, which therefore extends the application fields compared to rigid RFID sensor devices.

Commercial Solutions

Due to the great commercial opportunities in the RFID sensing techniques in the future IoT era, some pioneers in the industry including TI, STMicroelectronics, ASM, Farsen, Axzon, and Impinj have also devoted much effort to the research and development in this particular field. Some novel RF-to-DC converter modules, RFID transceivers, and RFID ICs are invented which can be found on the market. Some novel exemplar technical solutions for different applications are found on the market as well.

Promising RFID ICs for Sensor Development

Many RFID ICs were invented as RFID transponders in RFID applications, including both UHF EPC Class-1 G2 and NFC ISO/IEC14443 and ISO/IEC15693. Most of these ICs integrate RF energy harvesting circuits, internal logic control and memories, and a serial port to allow the read/write capability for an external MCU in order to update the sensor data. Analog-to-Digital Converters (ADCs) are included in some RFID ICs for interfacing with sensor components, such as the MLX90129 and SL13A. The model RF430FRL152H from TI integrating a low-power microcontroller MSP430 and a 14-bit digital signal A/D interface has effectively facilitated the further development. SL900A and Magnus-S3 M3D IC have gone further by integrating temperature and capacitive sensors, and integrating temperature, moisture sensors, and proximity sensors, respectively. The integration of built-in sensing modules allows even more convenient development for different applications. The most convenient choices for RFID sensor development are the Rocky100, RF430CL330H and RF430FRL152H, which have RF-to-DC converters and also power output in order to power the external sensors and MCUs. They can be used to develop battery-less RFID sensor devices.

Commercial RFID Sensor Applications

Based on the RFID ICs, there are also sample application modules. Farsens has produced a series of UHF RFID solutions for industrial, agricultural and logistics applications, including battery-less sensor modules, including temperature sensors, ambient light sensors, pressure sensors, magnetic field sensors, humidity/moisture sensors, force/

strain sensors, and RF field sensors. Similarly, On Semiconductor provides UHF RFID battery-free wireless sensors solutions based on the MagnusS2 Sensor IC, including moisture/proximity and temperature/proximity sensors. In addition, TI has provided some NFC sensing modules based on its RFID IC products, such as a NFC temperature sensor patch providing temperature readings to NFC enabled smartphones, and NFC keyboards as an alternative to touchscreen keypads for NFC-enabled smartphones and tablets. Novel developments are not limited to the abovementioned examples. Due to the features in wireless power and data transmission, battery-less and light-weight, contactless sensing, the RFID sensing techniques will play an important role in the future IoT market, including consumer electronics, manufacturing, agriculture, medical devices and logistics, etc.

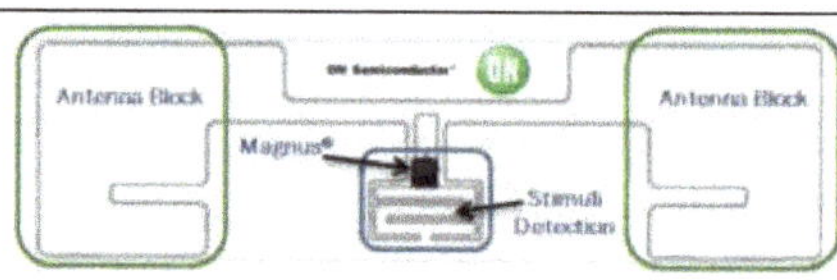

(a) Battery-less UHF moisture/proximity sensor

(c) Battery-less NFC temperature sensor patch

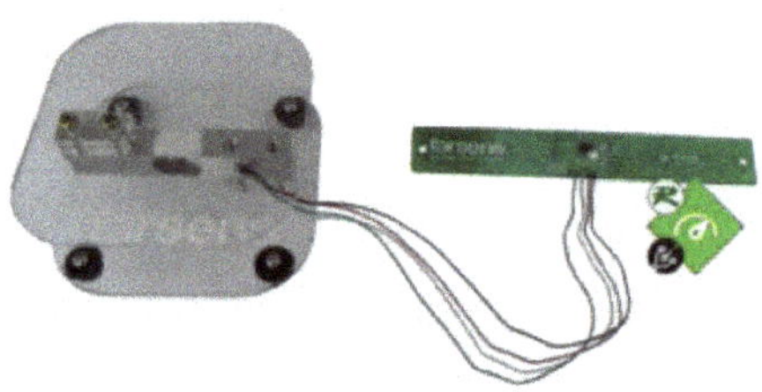

(b) Battery-less UHF force/strain sensor

(d) NFC Keyboard

Innovative Applications

RFID sensor techniques, taking advantage of both wireless power and data transfer and object identification, have been a new way of sensing and communication which can now find applications in a wide spectrum of fields. On the one hand, the wireless power and data transfer provides an interface for transmitting sensing data out from the object under test without wires and batteries, which results in substantial miniaturization of the sensor devices. On the other hand, the unique identification code distinguishes the object under tests and integrates the data with its ID directly. In many aspects, these two strengths make RFID sensing devices more competitive than traditional wireless sensor nodes, which need to carry a battery and a wireless module. Therefore, with the novel integration of RFID and lightweight sensor techniques, some measurement tasks which are technically challenging in the past become much more convenient. The fields of application include consumer electronics, healthcare, food and agriculture, chemical engineering, manufacturing and logistics, civil engineering, automotive, machinery, etc., which are summarized with typical examples as shown in figure.

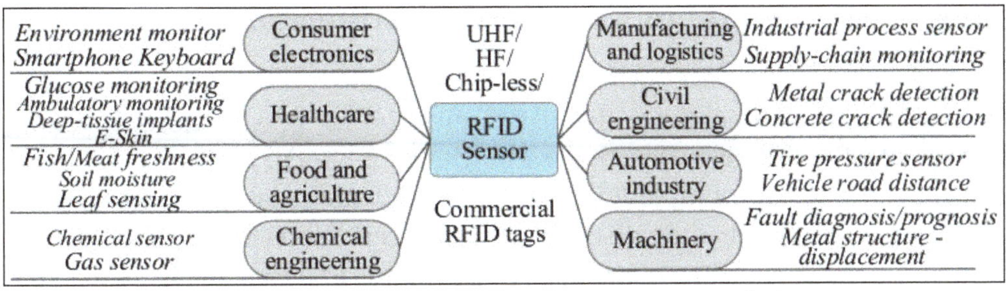

Application Fields and Examples of RFID Sensor Techniques.

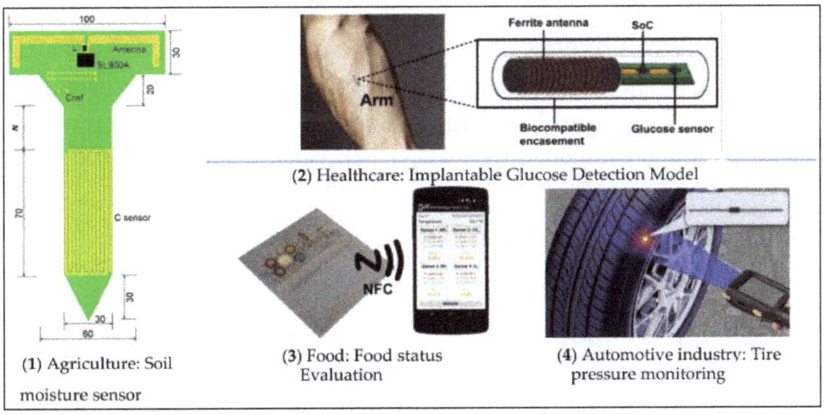

(1) Agriculture: Soil moisture sensor (2) Healthcare: Implantable Glucose Detection Model (3) Food: Food status Evaluation (4) Automotive industry: Tire pressure monitoring

Typical Application of RFID Sensors.

- RFID sensors of different types, including HF, UHF, and chip-less RFID devices have been widely used and integrated with cutting-edge technologies in different fields in recent years. This has been an effective and cost-efficient means for accessing sensing data in a battery-less, wireless, and passive way, which were significant technical challenges before RFID was invented.
- The RFID technique, as a power and communication interface in sensor solutions, is easy to integrate with other sensing modules with different processes, including inkjet printing, CMOS, flexible PCB, etc. Therefore, the functionality is easily extended for different purposes and application scenarios.
- The sensor techniques are integrated with RFID in different ways: a. Digital sensors with RFID ICs; b. Sensitive materials integrated with RFID antennas; c. Commercial RFID tags; d. Chip-less RFID antenna as a sensor.

Technical Challenges

Although continuous technical progresses have been made and many innovative applications of RFID-based sensor techniques are found, most of the presented works are still in the stage of proof-of-concept demonstration and in-lab test and evaluation. It is hard to find RFID sensor applications on the market besides a few commercial solutions. There is still a big gap between the investigations and practical applications. In summary, the key technical challenges are identified and listed as follows.

Efficiency of RF Frontend Energy Harvesting and Power Conversion

For an RFID tag, the antenna and IC just receive the RF power, store the energy and power the controller to respond to the read/write commands. All the power conversion and logical operations are completed in the RFID IC, which makes energy harvesting efficient, and therefore the reading distance normally reaches as far as 10 m. However, for RFID sensors, the RF energy harvesting front end needs to power the RFID IC, the MCU, and the sensing module. Although all the components are power efficient, the operation logic of the sensors are more complex and time consuming. Therefore, it is still a challenge to power all the components and cover the logic operations with RF energy harvesting only. Especially when the sensors are implanted in the materials under test, the RF signal is attenuated by the ambient materials and the received RF energy can hardly power all the operations, which seriously affects the read/write range of the RFID sensor.

Heterogeneity in the Enabling Techniques

The underlying techniques of RFID sensors show great heterogeneities in antennas, IC functionalities, sensing components, and data protocols, etc. The heterogeneity hinders the techniques from cross-platform integration and standardization. Most designs in the investigations are custom designed for the particular sensing scenarios, including the antennas, ICs, control logic, and data transmission. On the one hand, in order to gain high energy harvesting efficiency, the impedance of the antenna and RFID IC ports should satisfy a conjugate matching. Due to the variations of the impedance in RFID IC ports, the antennas need to be custom-designed for different ICs. For example, the port impedance for Impinj Monza 4 at 915MHz is $(11 + j143)\Omega$ and those for the AMS AG SL900A and Farsens Rocky100 at 915 MHz are $(123 - j303)\Omega$ and $(52 - j479)\ \Omega$, respectively, which results in a requirement for custom-designed antennas. On the other hand, the sensor data transmission protocols are variables for different applications. UHF sensor solutions may utilize the user ID section to accommodate the sensor data. But there are no widely accepted coding protocols for different sensor categories and measurement accuracies.

Reliability

RFID sensors are mainly attached to the measured object for identification and parameter sensing. However, they are also influenced by certain environmental factors in their actual applications. This not only requires high performance RFID sensors, but also requires RFID sensors to have the ability to cope with harsh environmental features such as high temperature, high pressure, humidity and impact. The metal package also blocks the information transmitted by the RFID sensor, causing the RFID sensor to malfunction. Therefore, in order to obtain stable RFID sensor information, the reliability of RFID sensor devices and antennas under severe conditions still poses great challenges.

The above technical issues have been the dominant obstacles facing the progress of RFID sensors. However, the products of the pioneer companies such as TI, AMS AG, Impinj, Farsens, Axzon, and some innovative applications have paved the way for building blocks for RFID sensor techniques. By overcoming the technical issues in energy harvesting efficiency and heterogeneities, the RFID sensing techniques will play an important role in the future rich sensing IoT world.

References

- Radio-Frequency-Identification-RFID-315656955: researchgate.net, Retrieved 22 May, 2020
- Characterization-of-the-identification-process-in-rfid-systems, radio-frequency-identification-fundamentals-and-applications-bringing-research-to-practice: intechopen.com, Retrieved 13 April, 2020
- High-frequency-hf-rfid-tags-systems, community: everythingrf.com, Retrieved 13 March, 2020
- Distance-sensitive-High-Frequency-RFID-Systems-224385310: researchgate.net, Retrieved 16 July, 2020
- Ultra-high-frequency-uhf-rfid-tags-systems, community: everythingrf.com, Retrieved 29 April, 2020

2

Radio Frequency Identification Tags

An electronic tag that exchanges data with an RFID reader through radio waves is referred to as a radio frequency identification tag. It includes active tags, passive tags, clipped tag, bag tag, organic RFID tags, etc. This chapter delves into the subject of radio frequency identification tags to provide a thorough understanding of it.

RFID tags are a type of tracking system that uses smart barcodes in order to identify items. RFID is short for "radio frequency identification," and as such, RFID tags utilize radio frequency technology. These radio waves transmit data from the tag to a reader, which then transmits the information to an RFID computer program. RFID tags are frequently used for merchandise, but they can also be used to track vehicles, pets, and even patients with Alzheimer's disease. An RFID tag may also be called an RFID chip.

Working of RFID Tags

An RFID tag works by transmitting and receiving information via an antenna and a microchip — also sometimes called an integrated circuit or IC. The microchip on an RFID reader is written with whatever information the user wants.

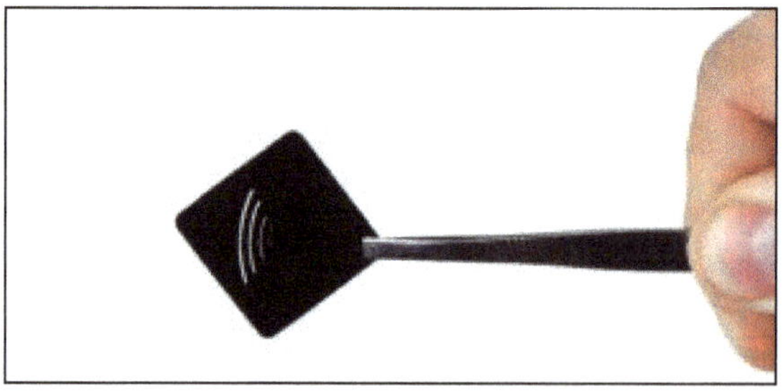

There are two main types of RFID tags: battery-operated and passive. As the name suggests, battery-operated RFID tags contain an onboard battery as a power supply, whereas a passive RFID tag does not, instead working by using electromagnetic energy transmitted from an RFID reader. Battery-operated RFID tags might also be called active RFID tags.

Passive RFID tags use three main frequencies to transmit information: 125 – 134 KHz, also known as Low Frequency (LF), 13.56 MHz, also known as High Frequency (HF) and Near-Field Communication (NFC), and 865 – 960 MHz, also known as Ultra High Frequency (UHF). The frequency used affects the tag's range. When a passive RFID tag is scanned by a reader, the reader transmits energy to the tag which powers it enough for the chip and antenna to relay information back to the reader. The reader then transmits this information back to an RFID computer program for interpretation. There are two main types of passive RFID tags: inlays and hard tags. Inlays are typically quite thin and can be stuck on various materials, whereas hard tags are just as the name suggests, made of a hard, durable material such as plastic or metal.

Active RFID tags use one of two main frequencies — either 433 MHz or 915 MHz — to transmit information. They contain three main parts, including a tag, antenna, and interrogator. The battery in an active RFID tag should supply enough power to last for 3-5 years. When it dies, the unit will need replaced, as the batteries are not currently replaceable. There are two main kinds of active RFID tags: beacons and transponders. Beacons send out an information ping every few seconds, and their signal is readable from several hundreds of feet away. Because they are sending out data so frequently, their battery tends to deplete quicker. Like passive RFID tags, transponders require the use of a reader to transmit information. When within range of one another, a reader first sends out a signal to the transponder, which then pings back with the relevant information. Because they only activate when near a reader, transponders are much more battery-efficient than beacons.

Examples of RFID Tags

Since an active RFID is constantly sending out a signal, it makes an excellent choice for those looking for up-to-the-minute live tracking, such as in tolling and real-time vehicle tracking applications. They are an expensive product, but they do offer a long read range, which may be preferred depending on their application.

Passive RFID tags are a much more economical choice than active RFID tags, and cost around 20 cents each. This makes them a popular choice for supply chain management, race tracking, file management, and access control applications. While a passive RFID tag does not require a direct line of sight to the RFID reader, it has a much shorter read range than an active RFID tag. They are small in size, lightweight, and can potentially last a lifetime.

Since active RFID tags feature a larger, more rugged design than passive RFID tags, they are better suited for applications where durability is required. They are frequently used in toll payment transponder systems, cargo tracking applications, and even in devices used to track people.

RFID Class-1 Generation-2 ("Gen-2") tags are used for item level identification in retail environments. "Class-1" refers to the functionality of the tag while "Gen-2" refers to the physcial and logical standards of tag and the encompassing system. These standards are maintained by EPCglobal. “Gen-2” standards were adopted in December of 2004 and offer many advantages over “Gen-1” such as: – An interoperable, global standard – Faster and more flexible read speeds – Faster, more accurate performance through the use of advanced anticollision protocols – An easier way to deploy many readers at one time to end users – Enhanced sercuirty and privacy.

RFID tags are classified as Class 0 through Class 5, depending on their functionality:

Class 0	UHFl read-only, preprogrammed passive tag.
Class 1	UHF or HF; write once, read many (WORM).
Class 2	Passive read-write tags that can be written to at any point in the supply chain.
Class 3	Read-write with onboard sensors capable of recording parameters like temperature, pressure, and motion; can be semipassive or active.
Class 4	Read-write active tags with integrated transmitters; can communicate with other tags and readers.
Class 5	Similar to Class 4 tags but with additional functionality; can provide power to other tags and communicate with devices other than readers.

RFID TAGGING

RFID tagging uses small radio frequency identification devices to track and identify objects. An RFID tagging system includes the tag itself, also known as a transponder; a scanning antenna and receiver, often combined into one reader, also known as an interrogator; and a host system application for data collection, processing and transmission.

Such systems are commonly used for the tracking and management of goods, animals and humans.

Uses and Applications

The most common RFID application is to track and manage goods and people in industries ranging from healthcare and industrial to manufacturing, retail and business. These applications include:

- Access control.
- Asset tracking.
- Counterfeit prevention.
- Document tracking.
- ID badging.

- Inventory management.
- Logistics.
- Personnel tracking.
- Supply chain management.

RFID is widely used in the supply chain as an alternative to barcode technology. Although more expensive than barcode stickers, RFID tags don't get dirty or fall off, and they don't require an unobstructed line of sight between the tag and reader.

Injectable RFID tags have been used to track wildlife and livestock, and even for human use. For example, injectable RFID tags can help medical personnel identify a patient who is unable to speak and even provide access to the person's medical records.

Various RFID tags.

Design

An RFID system contains a transponder, interrogator and the host applications. RFID systems generally provide one-way communication, with a transponder transmitting data to the interrogator.

Transponders, or the RFID tags, consist of a microchip, memory and antenna. Active RFID tags have their own power source (often battery), while passive tags have no power source, and are activated by the RF scan of the interrogator. As a result, active tags have more memory and can generally be read at greater ranges. In semi-passive RFID tags, a battery runs the circuitry while communication is powered by the RFID reader. Tags can be read-only or have read-write capabilities.

Smart labels, or smart tags, have emerged as a category of specially configured flat RFID tag that can be inserted into an item identification slip. Smart labels can be used to detect parameters in shipped goods, including temperature, shock, tilt, vibration and more.

Printable RFID tags use a low-power integrated circuit (IC) with an antenna and can be printed on demand from RFID smart label printers/encoders that embed the tag into the label material. Depending on the application, RFID tags can cost less than 10 cents or up to $50 apiece.

RFID readers, or interrogators, are the devices that gather data from the tags. RFID readers are network-connected and can be permanently attached or portable devices. They use RF waves to transmit signals to activate RFID tags. Once activated, the tag sends a wave back to the reader, where it is translated. RFID tags do not have to be scanned directly and do not require line of sight with an interrogator. The range at which a reader can transmit with a tag depends on the frequency used.

RFID host computers or networks are where the collected data is read, processed and passed onto the appropriate application for analysis and further use. Many RFID readers also contain onboard processing, running applications rather than relying on a host computer, or store the data until it can be uploaded to the host.

Challenges

RFID tagging is somewhat controversial, as tags can theoretically be cloned or used for illicit tracking. Another concern is that RFID tag data can be read by anyone with a compatible reader. Additionally, tags can sometimes be read after an item leaves a store or the supply chain. Tags can also be read without a user's knowledge, and it if has a unique serial number, it can be associated to a particular person. This is a privacy concern for individuals, while in military or medical scenarios it can be a national security issue or life-or-death matter.

Unfortunately, because RFID tags don't have a lot of compute power, they are unable to accommodate encryption. One exception to this is the RFID tags used in passports, basic access control (BAC). These chips have sufficient compute power to decode an encrypted token from a reader to validate the reader, and as a result, share the information in the tag. This is a sophisticated use case and often cost-prohibitive for other applications, however.

Beyond security and privacy, RFID tagging systems have two main issues: reader collision and tag collision. Reader collision - when a signal from one RFID reader interferes with a second reader - can be prevented by using an anti-collision protocol to make RFID tags take turns transmitting to their appropriate reader. Tag collision occurs when too many tags confuse an RFID reader by transmitting data at the same time. Choosing a reader that gathers tag info one at a time can prevent this issue.

RFID Standards and Regulations

RFID technology has a number of guidelines and specifications, the main ones being the International Organization for Standardization (ISO), Electronics Product Code Global Incorporated (EPCglobal) and International Electrotechnical Commission (IEC).

Each radio frequency has associated standards, including ISO 14223 and ISO/IEC 18000-2 for low-frequency RFID, ISO 15693 and ISO/IEC 14443 for high-frequency RFID, and ISO 18000-6C for ultra-high-frequency RFID.

Regulations may include the frequency at which authorized RFID applications can run in certain frequency ranges, maximum wattage allowed, reader-to-tag communication techniques, and channels/channel spacing, which involves how a frequency range is divided and involves preventing reader-to-reader interference.

RFID Alternatives

The use of RFID tagging as an alternative to barcodes is increasing. While barcodes require direct line of sight and closer proximity than an RFID tag, they also take longer to read, are not read-write and represent a product type versus the individual product represented by an RFID tag.

Barcodes are a much cheaper alternative to RFID tags, however. While different from a standard barcode, quick response (QR) codes are also cheaper than RFID codes, but they require direct line of sight and may not be suitable for all RFID applications.

Near-field communication and RFID are very similar and used for similar applications, but NFC offers two-way communication, enabling card emulation and peer-to-peer sharing. And while NFC operates at the same frequency as high-frequency RFID, NFC-enabled devices must be closer in proximity as NFC has a shorter read range. NFC is often used for access control and contactless payments, among other uses.

Bluetooth Low Energy (BLE) tags are similar to active RFID tags, as they have built-in batteries and offer long read ranges, but they also offer two-way communication, unlike most RFID. BLE-enabled tags are increasingly viewed as a more energy efficient alternative to RFID. BLE tags are more costly than passive RFID tags, however, though pricing is on par with some active RFID tags.

PASSIVE TAGS

Passive RFID systems use tags with no internal power source and instead are powered by the electromagnetic energy transmitted from an RFID reader. Passive RFID tags are used for applications such as access control, file tracking, race timing, supply chain management, smart labels, and more. The lower price point per tag makes employing passive RFID systems economical for many industries.

Active RFID systems use battery-powered RFID tags that continuously broadcast their own signal. Active RFID tags are commonly used as "beacons" to accurately track the real-time location of assets or in high-speed environments such as tolling. Active tags provide a much longer read range than passive tags, but they are also much more expensive.

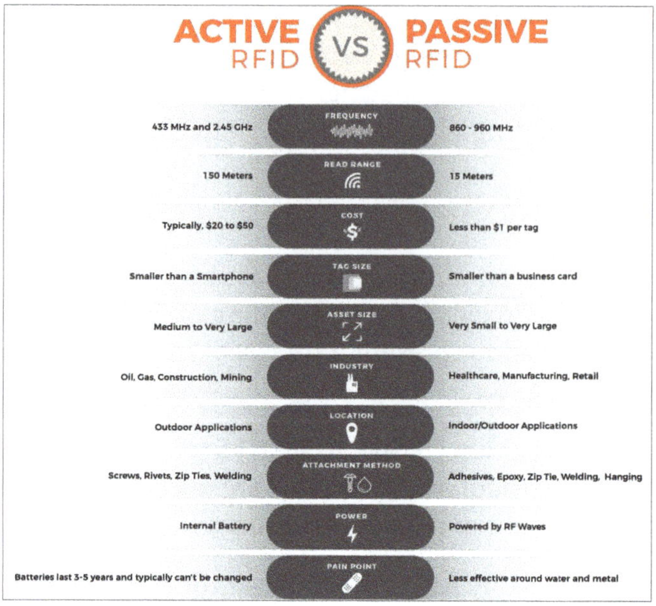

Passive RFID

Generally speaking, three main parts make up in a passive RFID system – an RFID reader or interrogator, an RFID antenna, and RFID tags. Unlike active RFID tags, passive RFID tags only have two main components – the tag's antenna and the microchip or integrated circuit (IC).

As the name implies, passive tags wait for a signal from an RFID reader. The reader sends energy to an antenna which converts that energy into an RF wave that is sent into the read zone. Once the tag is read within the read zone, the RFID tag's internal antenna draws in energy from the RF waves. The energy moves from the tag's antenna to the IC and powers the chip which generates a signal back to the RF system. This is called backscatter. The backscatter, or change in the electromagnetic or RF wave, is detected by the reader (via the antenna), which interprets the information.

Passive RFID tags have no internal power source, and a standard passive RFID tag consists only of an IC and internal antenna; this basic structure is commonly referred to as an RFID inlay. Countless other types of passive RFID tags exist on the market, but all tags generally fall into two categories – inlays or hard tags. Hard RFID tags are durable and made of plastic, metal, ceramic and even rubber. They come in all shapes and sizes and are typically designed for a unique function, material, or application.

A few different groups work to further divide passive hard tags; however, some tags will exist within two or more groups.

- High Temperature: Certain industries, like healthcare, track the number of cycles that instruments undergo in punishing autoclaves. Specific passive RFID tags are designed to withstand extreme temperatures and accommodate for those types of applications, among others.

- Rugged: Applications in outdoor environments or tough warehouses need a tag that can withstand snow and ice, dust and debris, or even the crushing forces felt under a tractor wheel. For these applications, a highly rugged passive tag is needed to make the application successful.
- Size: Some applications have specific size constraints when tracking small or large items. Size is one of the more important questions to answer when choosing an RFID tag because there are many different sizes available.
- Materials: If an application requires tracking metal assets, UHF metal-mount tags may be the only option. These tags are specifically designed to mitigate the problems UHF RFID faces around metal.
- Embeddable: If tagging an item becomes a problem for specific applications due to significant wear and tear, embeddable tags can fit in small crevices and be covered in epoxy so the RFID tag is out of harm's way.

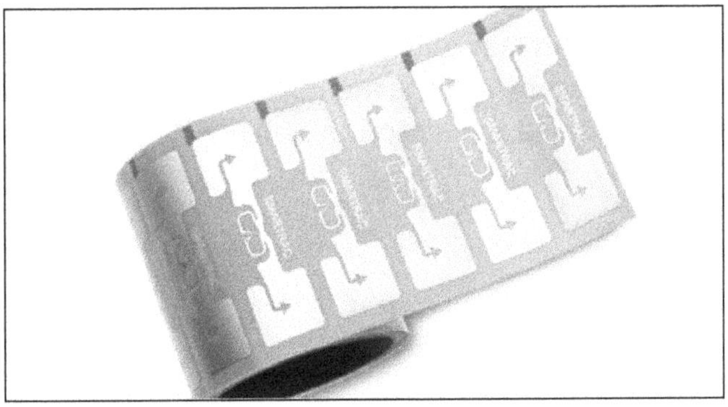

A roll of Passive RFID inlays.

Inlays are usually the cheapest RFID tags costing as low as $0.12 per tag in high volumes, but the price does not affect the performance. These inlays are grouped into three main types:

- Dry Inlays: An RFID microchip (IC) and antenna attached to a material or substrate called a web. These inlays look like they have been laminated and come standard with no adhesive.
- Wet Inlays: An RFID microchip (IC) and antenna attached to a material, usually PET or PVT, with an adhesive backing. Most of the time these inlays are clear and can be peeled off their roll and immediately stuck on an item.
- Paper Face Tags: These are essentially wet inlays with a white paper or poly face. These are ideal for applications that need printed numbers or logos on the front for identification.

Passive RFID tags do not all operate at the same frequency. There are three main frequencies within which passive RFID tags operate. The frequency range, along with other factors, strongly determines the read range, attachment materials, and application options.

- 125 – 134 KHz: Low Frequency (LF) – An extremely long wavelength with usually a short read range of about 1 – 10 centimetres. This frequency is typically used with animal tracking because it is not affected much by water or metal.
- 13.56 MHz: High Frequency (HF) & Near-Field Communication (NFC) – A medium wavelength with a typical read range of about 1 centimetre up to 1 meter. This frequency is used with data transmissions, access control applications, DVD kiosks, and passport security – applications that do not require a long read range.
- 865 – 960 MHz: Ultra High Frequency (UHF) – A short, high-energy wavelength of about a one meter which translates to long read range. Passive UHF tags can be read from an average distance of about 5 – 6 meters, but larger UHF tags can achieve up to 30+ meters of read range in ideal conditions. This frequency is typically used with race timing, IT asset tracking, file tracking, and laundry management as all these applications typically need more than a meter of read range.

As a general rule, higher frequencies will have shorter, higher-energy wavelengths and, in turn, longer read ranges. Moreover, the higher the frequency, generally speaking, the more issues an RFID system will have around non-RFID-friendly materials like water and metal.

Pros of Passive RFID:

- Smaller tags.
- Much cheaper tags.
- Thinner/more flexible tags.
- Higher range of tag options.
- Tags can last a lifetime without a battery (depending on the wear and tear).

Working of Passive RFID

How does Passive Tag Works?

In the figure, antenna creates electromagnetic fields thus a high frequency RF voltage produces. The voltage is rectified by a diode (a device which only allows current to flow in one direction), and the resulting signal is smoothed using a storage capacitor to create a more-or-less constant voltage that is then used to power the tag's logic circuitry and memory access. Passive tag memory circuitry is always non-volatile since the tag power is usually off. A similar rectification circuit, using a smaller capacitance value to allow the voltage to vary on the timescale of the reader data, is used to demodulate the information from the reader. This technique is known as envelope detection. Here we have shown a field-effect transistor (FET) used as a switch; when the FET is turned on, the antenna is grounded, allowing a large current to flow, and when it is off, the antenna floats allowing very little antenna current.

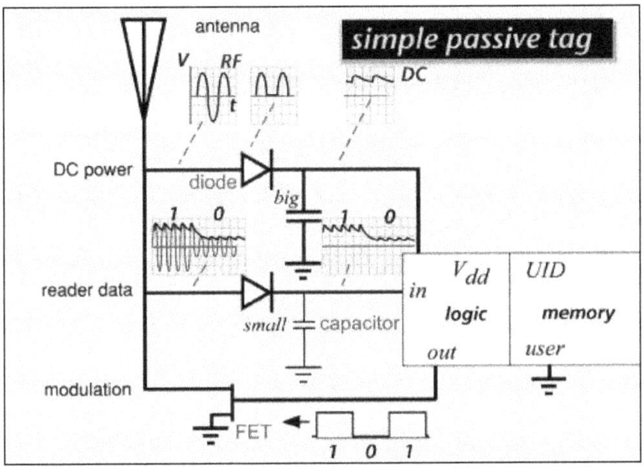

Schematic Depiction of Simple Passive RFID Tag.

RFID Modulation and Coding

Signals of interest for RFID are generally digitally modulated. A digitally modulated signal is a stream of distinct symbols. A simple example with substantial relevance for RFID is on–off keying (OOK).

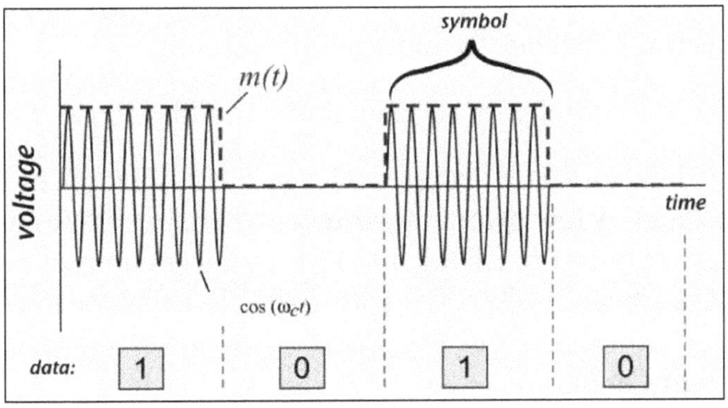

On–off keying (OOK).

A common solution to the power problem is to code the binary data prior to modulation. One RFID coding approach is known as pulse-interval encoding (PIE). A binary '1' is coded as a short power-off pulse following a long full-power interval, and a binary '0' is coded as a shorter full-power interval with the same power-off pulse.

Passive and semi-passive RFID tags use modulation of the reflected power from the tag antenna. In a passive RFID system, the transmitter does not turn itself off but instead, transmits CW during the time the receiver is listening for the tag signal. RFID radios use specialized components known as circulators or couplers to allow only reflected signals to get to the receiver, which might otherwise be saturated by the huge transmitted signal.

Description of simplified physics of backscatter signalling is given below.

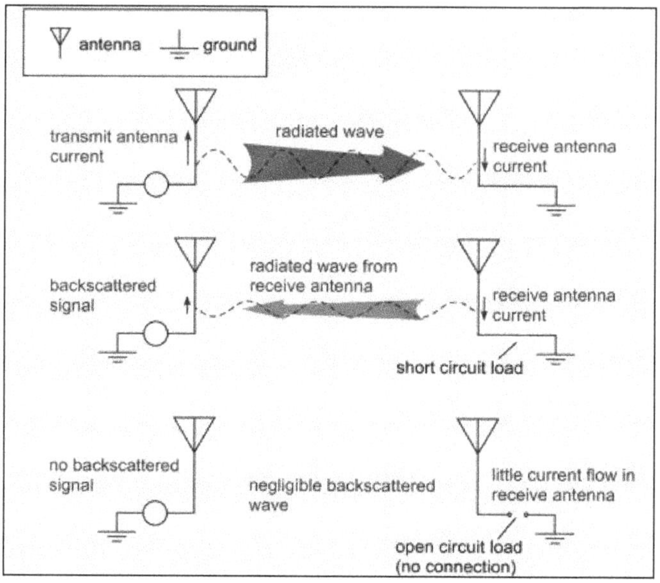

Simplified Physics of Backscatter Signalling.

Following are the benefits of backscattering Radio waves:

- Power supply to passive tag.
- Manage spectral efficiency to manage interference.
- The need for low complexity RF tags limits modulation and coding options.

It is important to provide power to Passive RF tag and detectable by th RF tag , so ASK is ultimate choice. There is few problem while we are using OOK and PSK. In OOK, It does not prove constant power to RF tag and PSK requires complex receiver design. But the benefits of ASK is , it can be detected with a simple envelope detector.

Reader Spectrum Control

The frequency bands available to RF tag system are divided into channels. 500Khz is the maximum channel. If the number of channels is greater than 50, the maximum conducted power is 1W. Interference from others readers can easily disrupt backscatter communication.

Radio communication is either near field or far field in pattern. There are different mechanism for EM wave communication. These are either electromagnetic or Magnetic in nature and depend on frequency.

Far-field (Electromagnetic)

Used in long distance and high frequencies (e.g UHF). Basically more sensitive to environment (e.g metal, water).

Near-field (Magnetic)

Used in short distance and low frequencies.

RFID Reader

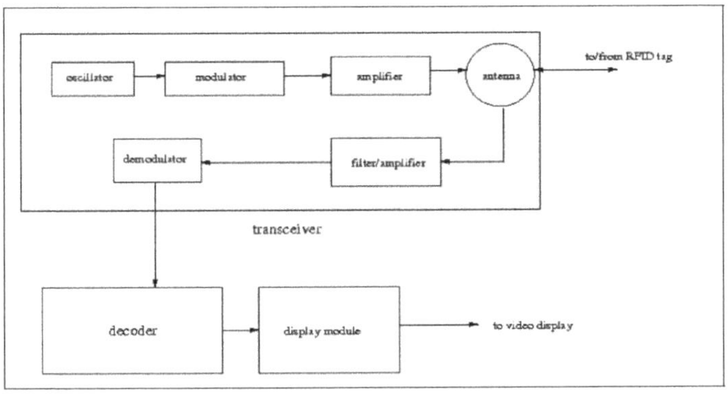

Block Diagram of RFID reader.

The above block diagram possesses three main subsystems. They are a Transceiver, a decoder and a display board. Antenna is used for communicating directly with the tag. A transmitter is used for interrogation of the tag and a receiver for the reply. RFID tags used amplitude modulated (AM). So the reader have to filter the signal for amplify. After this the reader demodulates the signal and obtain transmitted signal from tag.

A radio receiver needs to provide following properties such as Sensitivity, selectivity, dynamic range and flexibility.

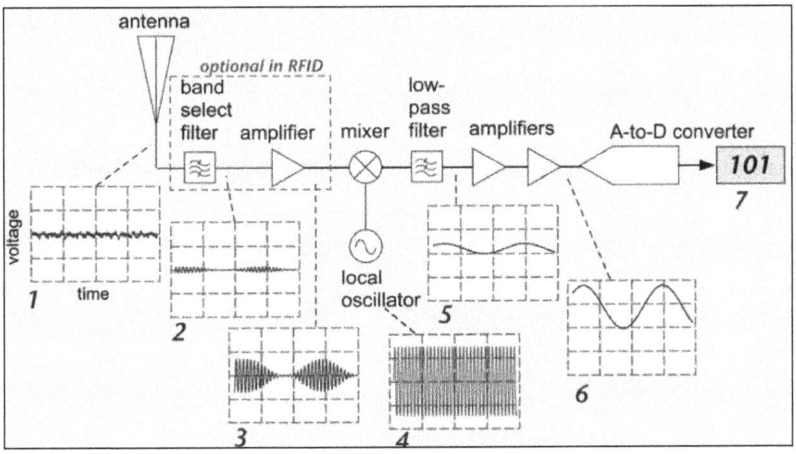

Example of Basic Circuitry.

All radios must perform certain generic functions. Small signals must be made larger (amplification), high-frequency signals must be generated (oscillation), signals at different frequencies must be combined to create new frequencies (mixing), and signals at wanted frequencies must be accepted while other frequencies are rejected (filtering). The components that perform these functions are unsurprisingly known as amplifiers, oscillators, mixers, and filters. In modern radios, signals must always be converted to and from digital form, though in RFID radios it can a bit grandiose to assign the terminology analogue-to-digital conversion to the simple comparators and switches that are often sufficient for these purposes.

UHF RFID Tags

In UHF RFID tag, it is important to take care about the duty cycle.

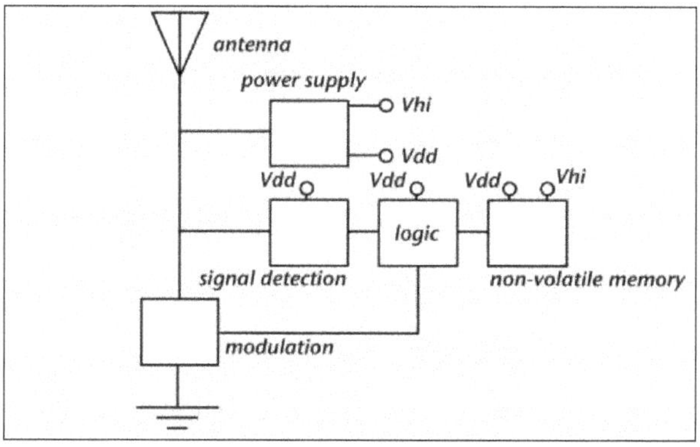

Elements of a passive UHF Tag.

In above figure, radar send the radio signal to the UHF tag. This signal captured by antenna and thus creates alternating current. From this current tag find both power and information. After retrieving all these signal tag sends data through antenna. By using back scattering signal returns to the reader.

For successful operation tag needs power. Tag seeks DC power.

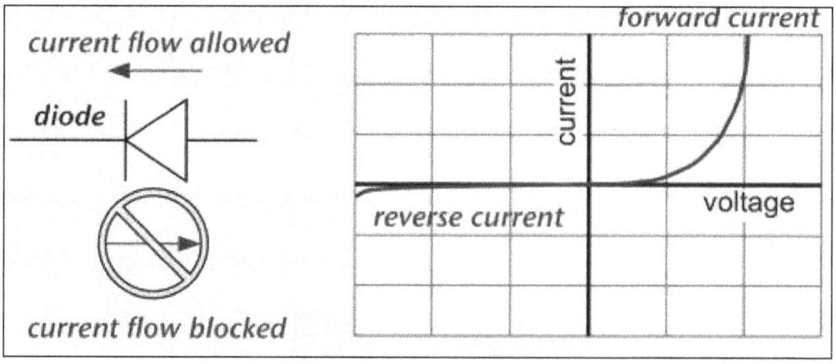

Diode working principle.

Using simple diode we can create alternating current to DC current.

The current flow through the diode is:

$$I = I_0\left(e^{\frac{qV}{kT}} - 1\right)$$

Where I is current, V is voltage, q is the charge and K is Boltzmann's constant. T is the temperature.

Rectifier Circuit in ON and OFF State

When the diode is in ON state, it means that the voltage across the diode is greater than the voltage required to turn ON the diode, the current starts flowing through the diode and then to capacitor and resistor. During ON state the capacitor stores the current and the current also flows through the diode.

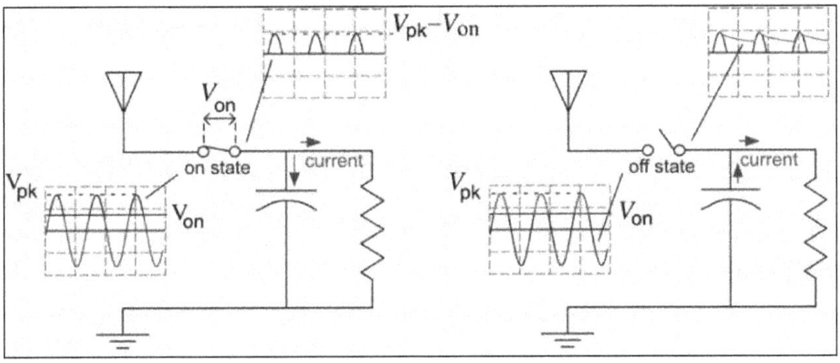

Rectifier circuit ON and OFF state.

During the OFF state the voltage across the diode is smaller than the turn on voltage, hence there will be no current across the diode. The capacitor starts discharging to the load. In this way we get constant RF signal to the tag.

Using rectifier circuits creates a problem. It cannot of turn on the IC due to required voltage. Charge pump combat against this problem. In the positive half cycle first diode is turned ON and in the negative half cycle other diode turned ON. In this way both the cycles provide enough charge to capacitor.

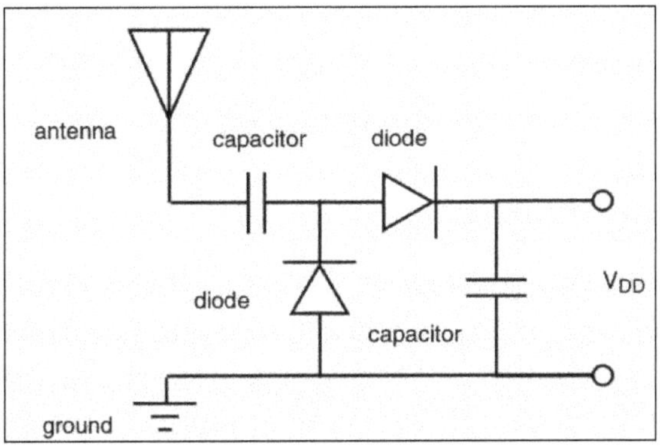

Voltage Doubler Schematic.

RFID Antennas

There are two types of antennas related to RFID. They are i) Reader Antenna and ii) Tag Antenna.

Reader Antenna

It is necessary to take care of the antenna properties like gain and radiation pattern, effective aperture, polarization, impedance and bandwidth.

Tag Antennas

Tag antennas are operates on same principle as reader antenna. But has some different challenges like cost, size , polarization, matching to the IC load. We will go through some important definition both common in Reader and Tag antenna.

Polarization

When the current flows, electric field is generated from the vector potential that is perpendicular to the direction of the current flow. But the tag antenna are long and thin and they do not behave like dipole and therefore the tag antenna will not receive any voltage and will not be read.

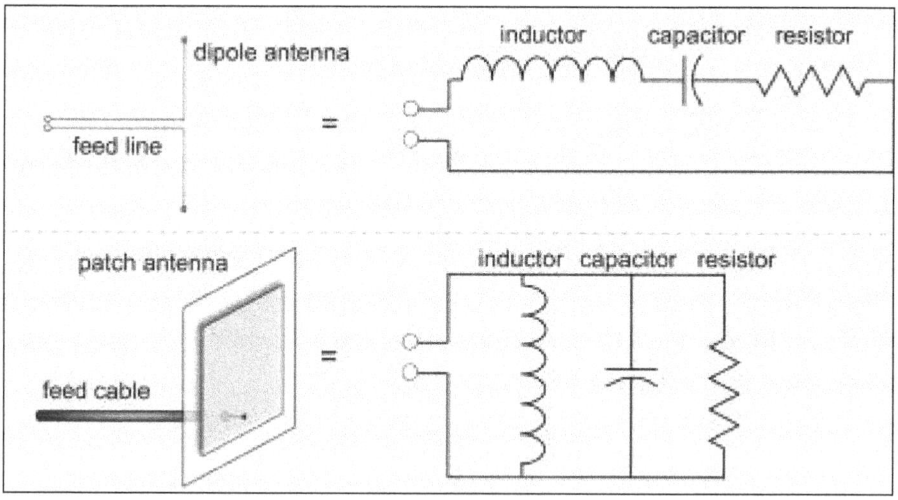

Types of Antennas.

Impedance and Bandwidth

An antenna carry resistance, capacitance and inductance so it is similar to RLC circuits. In the given diagram the upper figure is dipole antenna that is looks like a series combination of inductance, capacitance and resistance. At very low frequency the inductance has no effect and also capacitance stores small amount of charge. As frequency increases the inductance come into play and capacitance decreases and at resonance frequency both capacitance and inductance are equal in magnitude and have opposite signs, so the antenna behave like a pure resistance.

In patch antenna the structure is parallel combination of RLC than series combination. At resonant frequency both capacitor and inductor draws no current and therefore antenna behave like a resistance.

Impedance Matching and Power Transfer

The first goal of the tag antenna must be to deliver power to the tag IC to turn it on; if that doesn't happen, nothing else matters. So impedance matching and power transfer is necessary.

Another important criteria for antenna is dipoles and derivatives, Radar Cross section scattering, Near field and hybrid Tag Antennas.

Communication Protocol

An RFID communication protocol is like other communication protocol but they must response on the few matter like Medium, message format, medium access and context and interpretation.

Latest RFID communication protocols are i) EPC global Generation 1 ii) ISO 18000-6B (Intelligent) and iii) ISO 18000-6C (EPC global Class 1 Generation 2).

SEMI PASSIVE TAGS

The semi-passive tag operates similarly to the passive tag, using the reader signal to cause a response from the tag. The primary difference is that the semi-passive tag does have a battery - not for generating a response, but to power electronics that are used in conjunction with off-board sensors such as a thermal sensor. The sensor reading is incorporated into the tag return signal along with the tag serial number. Unsurprisingly, the semi-passive tag has most of the limitations noted for the passive tag in terms of slow read speeds and short read distances.

Issues in Integrated Circuit Design for UHF RFID

A passive UHF RFID tag, which has no battery on it, has to harvest energy from the electromagnetic waves which radiated from the reader and convert the RF power to DC supply to provide sufficient energy for the chip to function. The UHF RF signals have to travel a long distance and suffer a severe propagation loss in the air. Besides, to avoid harms to human health and the frequency interference, the emission power of reader is quite restricted according to local wireless regulations. Thus the RF power which finally reaches the tag is very low. If we further consider the multi-path effects and the environment influence to the tag antenna, the situation would be even worse. Therefore, the most important research topics in the passive UHF RFID tag design are to lower down power consumption of the tag chip and increase the efficiency of power harvesting.

Another research area of the tag is about anti-collision. Since the UHF RFID system has a higher data rate than the HF system, it provides the possibility for the reader to access an amount of the tags within a short time. If the tags are well organized by an effective anti-collision mechanism, a large data throughput can be achieved. Otherwise, the communication rate will be lowered down by the tag collision, and sometimes the communication may even fail.

As the RFID system plays an important role in information exchange, security issue is also a crucial consideration for designers. To achieve a lower cost, the area of the tag chip is required to be very small. This area limitation causes the encryption mechanism for security very hard to be implemented on the tag chip.

The cost of the tag is determined mainly by the package of the tag chip. The cost for bonding chip and fabricating antenna of tag now becomes the bottleneck to decrease the total cost. Several new techniques including on-chip antenna (OCA) have been developed to reduce the package cost.

Low Power Design

Apparently, one direct way to increase the operation distance of the tag is to reduce the power consumption of the tag chip. All the low power techniques used in other applications can also be used in tag design.

Digital

In an RFID tag, the digital circuit plays an important role as a central controller. To reduce the power dissipation of the digital part, one way is to lower the supply voltage. Since the internal clock frequency of the tag is quite low (a few megahertz), the latency of the digital circuit is not an issue here, leaving sufficient margin to reduce the supply voltage. The transponder IC in has a 1.5 V supply voltage in 0.5 m CMOS process. The operation voltage of the digital base-band circuits could be lowered to 1.14 V in 0.18 m CMOS process. Also in 0.18 m CMOS process, implements a digital core with a 0.6 V supply voltage based on the special low-voltage low-power standard cells design.

Another effort to reduce the power dissipation of the digital circuit is to lower the clock frequency. Most of the commercial UHF RFID protocols use PIE code or Manchester code in the communication link from reader to tag. Both of them require an on-chip oscillator to provide the sampling clock for the decoder and the operation clock for the digital core. The sampling clock is required to be much higher than the data rate in order to achieve better BER. In this case, most low power RFID designs adopt clock separation technique. For instance, a 3.35~3.75 MHz clock for PIE decoding and an 8 times slower clock for other components. Similarly, a digital clock manager is designed in, which uses a clock of 1 MHz to produce a synchronous 40 kHz clock with the Manchester-coded data input, and all the other digital circuits are using the low frequency clock.

To make good use of the idle states of the different components in the tag is also an effective way to further reduce the power consumption. Techniques such as gated-clock can be easily used in the tag.

Analog

The analog front-end of the tag is also a large current consumer. Low power supply voltage results more difficulty on the design of analog part. Since different analog components have different design considerations, it is impossible to cover all the aspects of the low power analog circuits design.

A voltage regulator in the passive RFID tag is required to perform well as the input RF power varies more than 30 dB. Such a large dynamic range requires that the regulator should have a sufficient bandwidth to track the voltage variation. But there is a tradeoff between bandwidth and power consumption. An extremely low power voltage regulator which dissipates only 110 nA current has been presented in. The highlight of this work is the adoption of the dynamic bandwidth boosting to achieve a wide enough bandwidth while keeping current dissipation within 110 nA. The authors design the regulator into two stages. Stage one is primarily functioning when the chip operates in the far field and it sinks only limited current. Stage two can sink a large mount of current and it mainly operates in the near field.

Low power on-chip oscillator design is also demanding and the ring oscillator is widely used. The problem is the frequency variation along with the change of the supply voltage or the environment. A current starve ring oscillator which has less than 7% frequency variation with the supply voltage change between 1.5 V and 3.0 V and 90ć temperature change. The authors use two complementary biasing circuits to generate a temperature independent bias current to drive the ring oscillator. Low voltage current mirrors are also adopted to reduce the effect of supply voltage variation and lower minimum operation voltage.

Rectifier Design

Apart from cutting down power consumption, another effective way to maximize communication distance is to optimize the power utilization, which can be achieved by design of a high efficient rectifier.

As shown in figure, the most commonly used rectifier architecture is the Schottky diode based Dickson voltage multiplier, which was first published in. Its test results show that the efficiency of the rectifier can reach 18% at 1.5 A current consumption when active.

Based on this work, many further researches have been made to improve the performance of the rectifier.

A CMOS-only full-wave rectifier, which has been reported in, can raise the measured efficiency up to 36.6%. This result is the highest rectifier efficiency measured so far. The proposed rectifier diagram is shown in figure. In this architecture, internal threshold voltage cancellation circuits are used to avoid threshold voltage drop. And substituting PMOS transistors for NMOS transistors reduces the parasitic capacitances which degrade the power conversion efficiency.

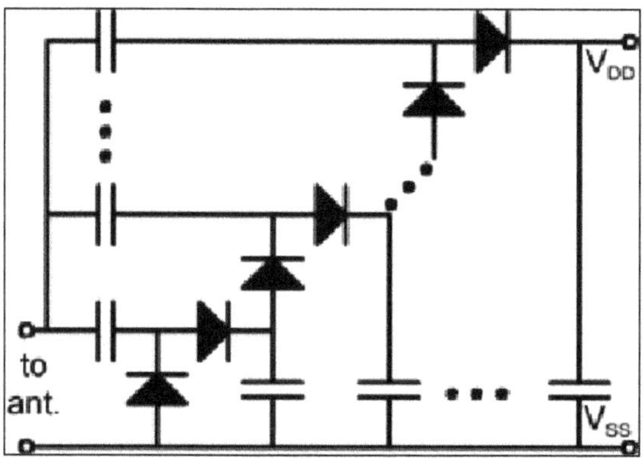

Schottky diode based Dickson voltage multiplier.

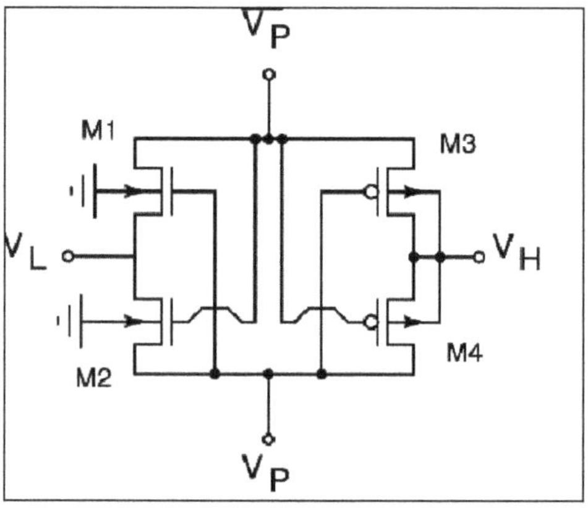

Four-transistor cell.

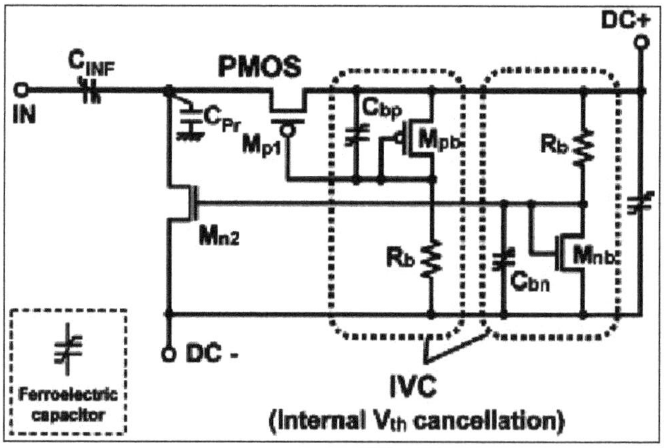

The proposed half-wave rectifier.

Anti-collision Mechanism

In a typical passive UHF RFID application, a reader is supposed to have the capability of reading more than one tag in the same field. When the reader sends a query command, all the tags in the field will answer the reader simultaneously. Then the tag collision which may fail the communication from tag to reader happens. Therefore, an anti-collision mechanism is essential in RFID system. And it has great effect on the throughput performance.

Actually, anti-collision mechanism is a system solution to the tag collision problem. Both reader and tag have to be involved in an anti-collision algorithm. Tag anti-collision protocols can be grouped into two broad categories, Aloha based protocols and tree based protocols. The principle of Aloha based protocols is to reduce the collision probability by separating tag transmission in distinct time slot. One key research issue in Aloha based method is to optimize the slot number while the number of the tags varies. The tree based protocols use a group splitting mechanism. Their main disadvantage is the relatively long identification delay. Therefore, a lot of works have been done to exploit an effective way to shorten identification time. An adaptive binary splitting protocol uses information obtained from the last process of tag identification to improve the efficiency. And creates a new binary search algorithm with interference cancellation. To minimize the anticollision cost, an enhanced binary search with cut-through operation. Although there are numbers of works in this area, anti-collision mechanism is still a hot research topic in RFID field.

Security Mechanism

The information security is always a stringent requirement in RFID systems. Due to the cost limitation, the conventional high security level encryption method is too hard to realize in the passive UHF RFID tags. There is a tradeoff between the hardware cost and the security level. And even more, the limited power consumption and the low clock frequency make the design of crypto module in a low cost passive RFID tag much more difficult.

Authentication and encryption are two major aspects in RFID security mechanism. So far, mutual three-pass authentication method is the most commonly used mechanism. The hardware implementation of proper encryption algorithm is one of the most challenging issues in tag design. A low power encryption hardware using TEA algorithm, which has an area of 0.21 mm2 and power of 7.37 W in 0.35 m CMOS process. An implementation of SHA-1 algorithm in 0.25 m CMOS technology has 10,641 gates and 19.5 W power consumption.

On-chip Antenna Design

Although the cost of the tag chip keeps decreasing with the development of the integrated circuit, it is still difficult to cut down the package cost of the RFID tag chip. In some applications, where the operation distance is a primary consideration, OCA can be an effective way to cut down the total cost and make the size of tag dramatically small.

Issues in Reader

A UHF RFID reader behaves like a wireless transceiver. The major difference between an RFID reader and other communication systems is that the RFID reader receives the signals backscattered from tags while transmitting continuous carrier waves of the same frequency. In other words, the receiver of an RFID reader suffers a carrier leakage problem because of the limited Tx-Rx isolation. Typically, the power of the leakage carrier is more than 0 dBm. Such a large leakage power easily saturates the RF front-end and then degrades the performance of the receiver. In this situation, the receiver needs to have sufficient linearity and an LNA is hard to be designed or used in receiver. To achieve better sensitivity, carrier leakage problem must be considered carefully.

The endless low cost demand requires that RFID reader chip should integrate as many components as possible. Design of a multi-mode, high efficient, high linearity power amplifier with sufficient output power is a great challenge in CMOS technology.

Carrier Leakage Problem

One way to alleviate the leakage saturation problem is to insert an attenuator before LNA, as shown in figure. By attenuating the input RF power, linearity of the following parts of the receiver can be released. But the insertion loss of the attenuator degrades NF severely because it is the first stage of the receiver. To solve the leakage problem while good NF and linearity performance is kept, a self-leakage canceller circuit is proposed by, as shown in figure. The canceller uses an amplitude tuner and a phase tuner to adaptively generate a counterpart of the leakage which has the same phase and amplitude. Then the leakage carrier can be cancelled by a differential amplifier. Figure depicts the principle of the proposed blocker rejecting LNA (BR-LNA). The BR-LNA consists of a linear LNA which amplifies both the leakage and the wanted signal and a non-linear limiter which only amplifies the large leakage signal.

CMOS PA Design

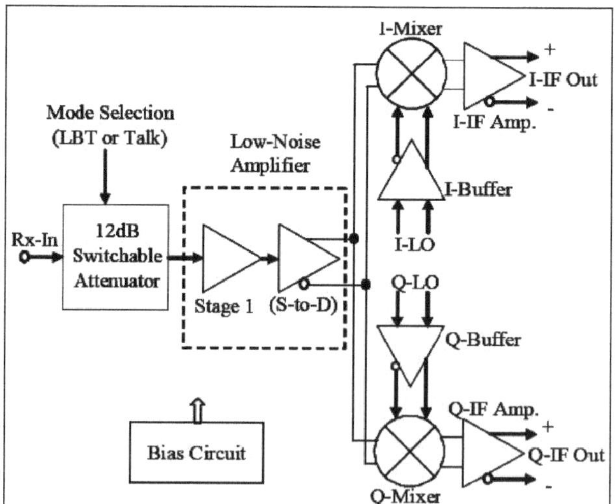

Switchable attenuator is inserted before the LNA.
The attenuator can be switched off in the LBT mode.

CMOS technology is the most popular fabrication process of commercial integrated circuit due to its low cost and relatively high performance. However, in the RF design area, the CMOS power amplifier is hard to output sufficient RF power with high efficiency and linearity compared to other processes, such as biCMOS.

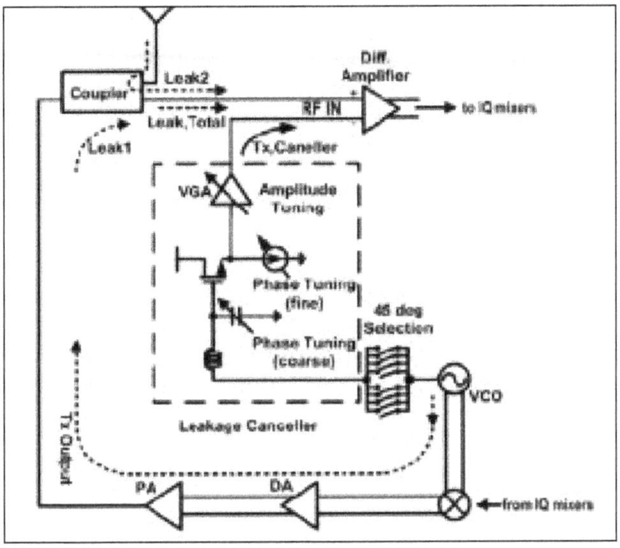

The proposed leakage canceller structure.

In single chip RFID reader, PA is required to have extremely high linearity since some of the widely used RFID protocols adopt DSB-ASK modulation. And the output power is supposed to be as large as possible to extend the operation distance. The power limitation also results more requirements on PA efficiency. Therefore, CMOS PA design is also a challenge in reader.

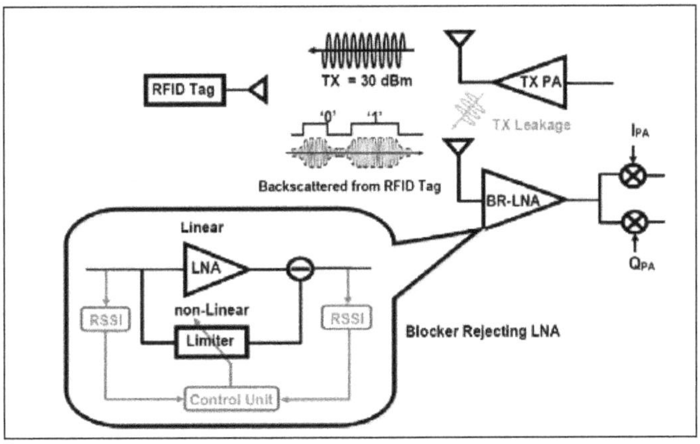

The principle of BR-LNA.

Design Instances

A Passive RFID Tag with Standard EEPROM

A 915 MHz passive RFID tag has been designed and fabricated in 0.18 m standard CMOS technology. The block diagram is shown in figure.

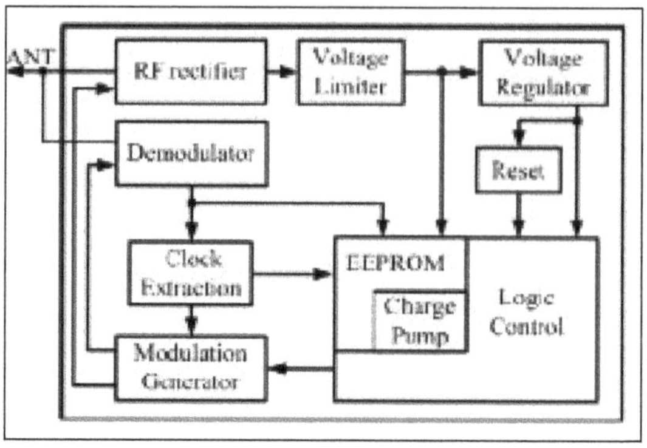

Block diagram of the designed 915 MHz passive RFID tag.

This chip is required to drive a standard commercial EEPROM which consumes 450 W power for reading and 6 mW for writing. Such a large power consumption of the load is the main challenge for RF front-end, especially for rectifier. And the voltage regulator also needs a careful design because of the large load current.

Supply voltage is generated by the Schottky diode based Dickson rectifier shown in figure. Although many efforts have been made to use MOS transistors as the rectifying devices, Schottky diodes are more suitable for large current applications. Figure shows a comparison of the DC character between a Schottky diode and a diodeconnected HV native NMOS. From this figure, we can see that both native NMOS and Schottky diode have almost the same forward-current driving capability, but the native NMOS should have a W/L more than 1000/1.8, which is 72 times larger than Schottky diode, and the source and drain will further enlarge the NMOS area. Therefore, the total area of the native NMOS is unacceptable. Moreover, when the area increases, the parasitic capacitance of the transistor and reverse leakage current will also increase. It will significantly degrade the efficiency of rectifier.

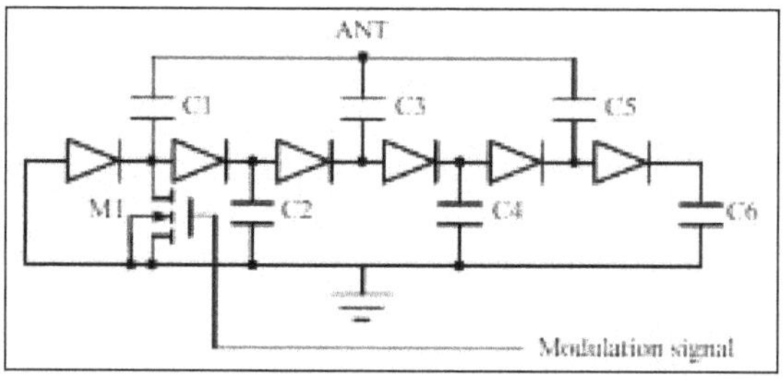

RF rectifier structure.

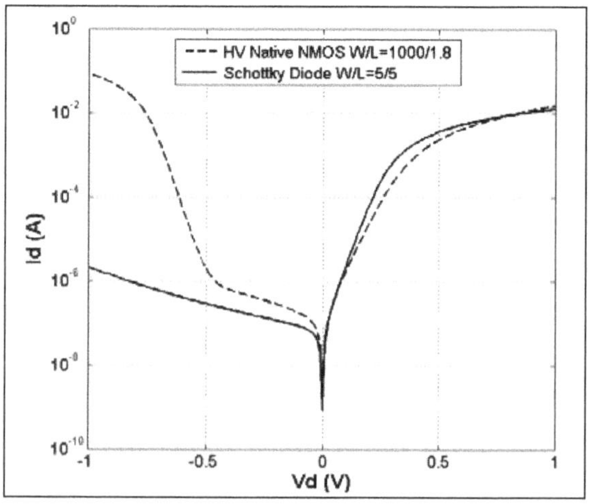

Comparison of DC character between HV native NMOS and Schottky diode.

To reduce power consumption, EEPROM is only turned on when there is a request to access memory. Frequently turn on and off the memory results in a large load current variation. Together with the change of the input RF power, it demands voltage regulator should have a sufficient bandwidth to tolerate variations. Figure shows the structure of the designed regulator. It consists of a voltage limiter and an LDO regulator. The voltage limiter uses four cascaded diodes as a current sink. It functions only when the input power is high enough. Since it is a simple shunt current structure, tracking the input voltage variation is not a problem. As the second stage of regulator, the LDO provides refined supply voltage. The gain of the amplifier is sacrificed to increase the bandwidth and keep current low, because the accuracy of the supply voltage compared to the reference is not the primary concern.

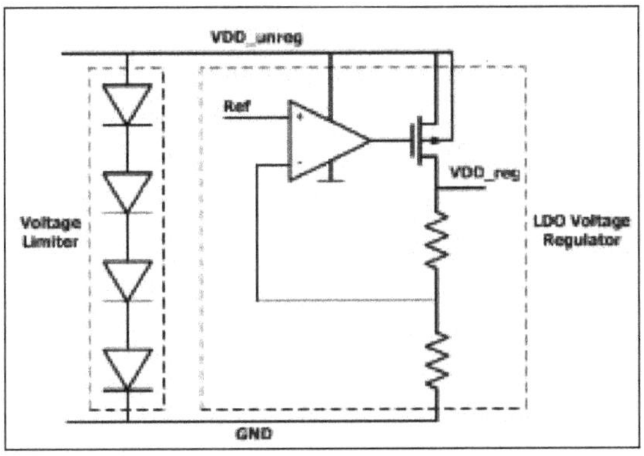

Structure of the regulator. The first stage is a limiter and the second is a LDO regulator.

A Security Module using XTEA Algorithm

In RFID security area, an XTEA based encryption module is implemented in a tag. To achieve low hardware cost and relatively high security level, XTEA algorithm is selected since it is simple to realize and has never been cryptanalyzed. Figure shows the working process of XTEA algorithm.

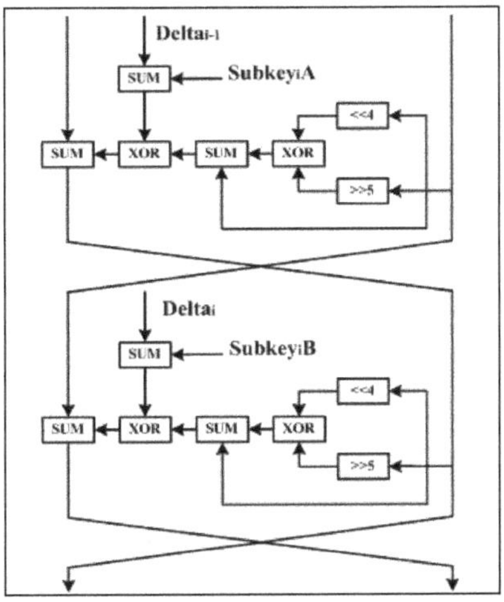

Working process of XTEA algorithm, cycle i.

The total logic gate number of digital core with the proposed encryption module is 4.7 K. Under a 40 KHz system clock, it takes only 17 ms to complete the whole authentication process, consuming 0.98 uW at 1.8 V power supply.

A Single Chip RFID Reader Transceiver

An 860~960 MHz single chip RFID reader transceiver is also under development. This transceiver adopts direct-conversion architecture and the block diagram of the transceiver is depicted in figure.

To cope with the carrier leakage problem, the received signals are directly down-converted into baseband using a simple passive mixer which has higher linearity than active mixers. The diagram of the mixer is shown in figure.

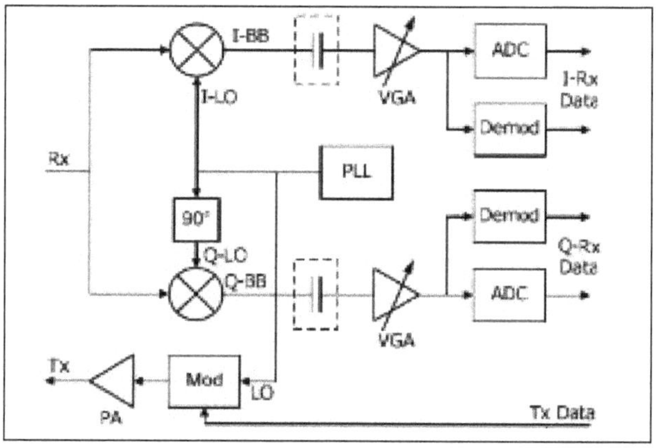

Block diagram of the architecture of the designed single chip RFID reader transceiver.

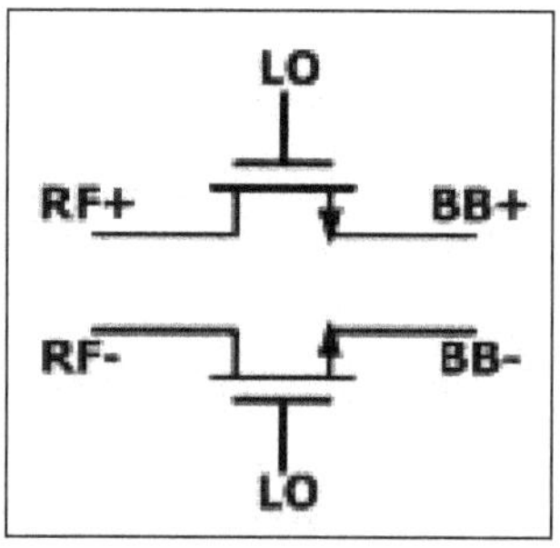

Block diagram of the passive high linearity mixter.

Two off-chip capacitances are used to filter out the DC components of the down-conversion mixers output. The variable gain amplifiers have a gain from 20 dB to 60 dB with 20 dB step. Two 2 Ms/s 14 bit ADCs are used to convert I-Q baseband signals into digital signals for further processing. The synthesizer has a frequency range from 860 MHz to 960 MHz with 250 kHz step.

The transmitter is composed of an ASK modulator and a CMOS PA. In order to obtain higher efficiency, a three-stage class-E PA is designed, shown in figure. The three stages are class-A, class-AB and class-E respectively. All the components are on chip except for the L2-C2 series resonator and quarter-wavelength transmission lines. A cascode structure is adopted in the output stage to enhance the output power and protect the transistors. Rf and Cf compose a feedback network to increases the stability of PA. Post simulation result shows that the output 1dB power is 25.1dBm with the PAE of 50.4%. The total power consumption of the chip is estimated no more than 500 mW when all the components are active.

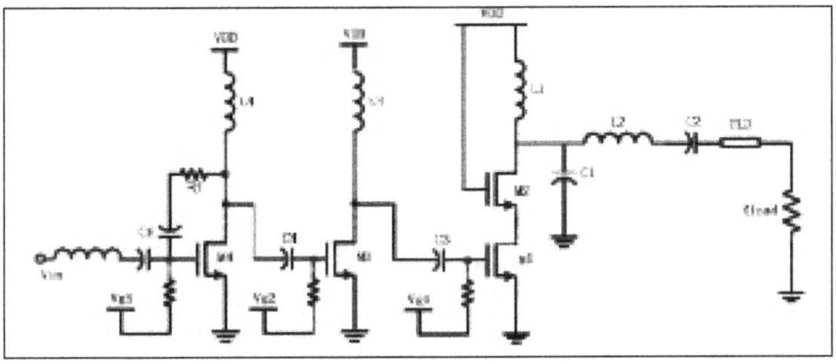

Proposed class-E power amplifier.

Dual-band UHF RFID Tag Antenna

A low profile dual-band passive Ultra High Frequency Radio Frequency Identification (UHFRFID) tag antenna designed to operate at two RFID bands allocated for use in Europe (865–868 MHz) and Japan (950–956 MHz) is proposed. The antenna has two eccentric circular rings of different radii to provide dual band response. An arc-shaped strip with Impinj Monza-4 IC chip is used to feed the two rings simultaneously by microstrip-line coupling-feed technique. The proposed design is simulated using Ansoft HFSS, and the prototype is fabricated. The return losses at 866 MHz and 952 MHz are measured to be−12.25 dB and−12.99 dB, respectively, which are in good agreement with the simulated results. The proposed antenna exhibits a 10 dB bandwidth of 9 MHz from 862 to 870 MHz and an 8 MHz 10 dB band width from 949 to 956 MHz covering the UHF RFID bands in Europe and Japan. The maximum read ranges are measured to be around 3 m in the 865–868 MHz band and 2.6 m in the 950–956 MHz band.

Of late, Radio Frequency Identification (RFID) system has turned out to be one of the most popular automatic identification techniques, replacing the omnipresent barcodes in an increasing number of cases. It has found widespread application in supply chain management, health care, inventory, animal tracking, and security. An RFID system consists of an electronic data-carrying device called the tag, an interrogator or reader, and a host computer with application software. Tags can be active (with on board source of power, like battery) or passive (without battery). A typical passive tag comprises an antenna and an IC chip. For practical applications, RFID tag antennas must be small, low profile, and suitable for low-cost mass production.

Generally four frequency bands are used for RFID systems; i.e., low frequency (125–134 kHz), high frequency (13.56 MHz), ultra-high frequency (860–960 MHz) and microwave frequency (2.45, 5.8 and24 GHz). The ultra-high frequency (UHF) tags have garnered more attention as they offer longer read ranges, higher reading speeds, enable simultaneous detection of more number of tags and need smaller antennas than low frequency (LF) and high frequency (HF) systems. Transfer of data between the tag and the reader is achieved by near-field coupling with load modulation in LF and HF bands, whereas, in UHF bands, the principle of backscatter modulation is used. In backscatter modulation, the tag antenna receives the modulated signal transmitted by the reader. The RF voltage developed across the antenna terminals powers up the chip. The chip acts as a switch to match or mismatch its internal load to the antenna by varying its input impedance leading to the effective modulation of the back scattered signal.

Different parts of the radio spectrum have been allocated by different countries for RFID, viz.866–869 MHz in Europe, 902–928 MHz in North and South America, and 950–956 MHz in Japan. One of the important challenges is to design tags which can be operated at any two or more of these regulated UHF bands simultaneously. The tags which are capable of operating at two of the allocated frequencies simultaneously are referred to as dual-band tags. The dual-band tags are proposed as an alternative to broadband tags, as they provide better matching at two specific bands and higher read ranges than the later. Several methods to obtain dual-band behaviour have been reported earlier in literature. One of the techniques is the simultaneous excitation of two orthogonal modes, i.e., TM01and TM10at the first and second resonant frequencies respectively by proper selection of the feed location. The use of multiple patches to generate dual bands is reported in. In these structures, each of the multiple patches of different resonant lengths supports strong currents and radiations at different resonant frequencies, hence providing dual-band behaviour. Dual-band response can also be obtained by introducing reactive loads, such as stubs, notches, and slots to a single patch, wherein the current distribution of one of the modes of the patch is modified to alter the resonant frequency of that particular mode without significantly affecting the other mode. Perturbation method to obtain dual-band response has been reported by Paredes et al. in. In this method, spiral or split ring resonators coupled to a mono-band antenna (resonating at an intermediate frequency f0) split the tag resonance into two (one beyond and the other below f0) by perturbing the line impedance and electrical length of the antenna. Some of other methods to achieve dual-band response involve the use of AMC ground plane in which the reflection phase of the ground plane is changed to obtain double resonance at two required frequencies of the material.

Antenna Design

The geometry of the proposed RFID tag antenna is shown in figure. It has two eccentric circular ring patches. An arc-shaped strip in between these two rings serves as the feeding strip and excites both the rings simultaneously. An eccentric circular ring patch covers smaller copper area than a regular circular patch. Therefore, it stores less energy beneath it than a circular patch of the same radius owing to its smaller dimension. This reduces the quality factor as the quality factor is directly related to the stored energy and is given by the equation:

$$Q = 2\pi \times \frac{\text{energy stored}}{\text{energy dissipated per cycle}}$$

With a decrease in the quality factor, the bandwidth increases as the quality factor is inversely proportional to the bandwidth. Thus, the eccentric circular rings lead to bandwidth enhancement of the tag antenna. The entire design is printed on a commercially available 1.6 mm thick, inexpensive,FR-4 substrate of dielectric constant $\varepsilon_r = 4.4$ and loss tangent tan δ=0.002. The overall size of the antenna is 91×91 mm2. The antenna design is simulated using Ansoft High Frequency Structure Simulator (HFSS) software. The optimized values of all the parameters are enlisted in table.

The microchips used in UHF RFID tags have capacitive input impedance. For maximum power transfer, tag antennas are designed to have inductive impedance for conjugate matching. In order to achieve dual-band response, the impedances of the tag antenna must be matched with the chip impedances at both the resonant frequencies. The arc-shaped short-circuited microstrip-line structure consisting of the microchip (at an angle $\beta = 22^{\circ}$ with the x-axis) is used to excite the two rings simultaneously by coupling feed technique. It has been reported in the literature that in coupling feed technique, the impedance of a tag antenna is contributed by the total length, width of the microstrip-line feeding structure and also the coupling distance between microstrip line and patch.

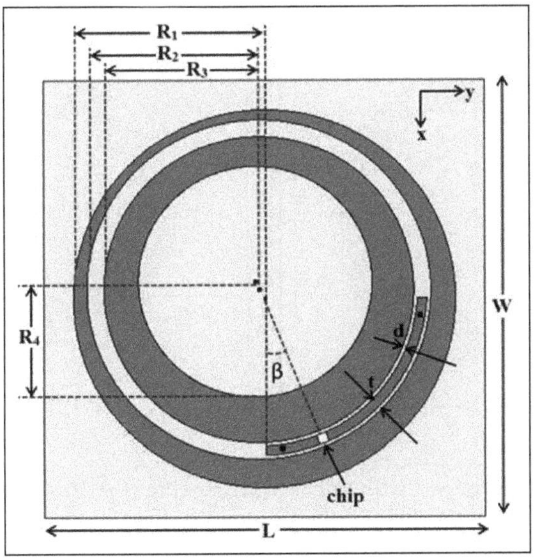

Design of the proposed antenna.

Table: Optimized values of the proposed tag antenna parameters.

Parameter	Values (mm)	Parameter	Values (mm)
L	91	r_1	59.9

W	91	r_2	46
R_1	39.4	r_3	26.6
R_2	35.4	r_4	23.4
R_3	31.9	t	2
R_4	24	d	0.75

The reactive part of the tag antenna impedance is mainly determined by the inductive reactance of the short-circuited microstrip line. To acquire an inductive reactance of X_c, the required length (l)of the arc-shaped short-circuited microstrip line can be calculated using the formula:

$$l=\frac{\lambda}{2\pi}\tan^{-1}\left(\frac{X_c}{Z_o}\right)$$

Where λ is the free space wavelength at the operating frequency, and Z_o is the characteristic impedance of the short-circuited microstrip line. The value of Z_o of a microstrip line is approximated as follows:

$$Z_o=\frac{60}{\sqrt{\varepsilon_e}}\ln\left(\frac{8h}{t}+\frac{t}{4h}\right)\text{ for }\frac{t}{h}\le 1$$

Where t is the width, and ε_e is the effective dielectric constant of the microstrip line.

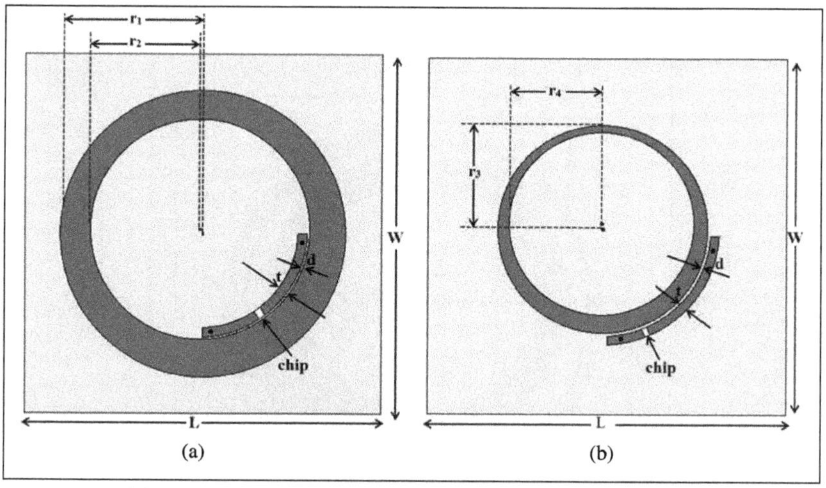

(a) Antenna 1 with only the larger ring (for 866 MHz).
(b) Antenna 2 with only the smaller ring (for 956 MHz).

Using equation above, the characteristic impedance Z_o is calculated to be 63.6 Ω for the arc-shaped strip of thickness t=2 mm. As the arc-shaped microstrip-line excites both the rings simultaneously, it is designed to obtain an impedance matching at an intermediate frequency of 911 MHz. By extra polation, the chip impedance at 911 MHz is obtained to be around 11.16–j143.65Ω. To obtain the reactance of j143.65Ω, we have chosen the total length of the arc-shaped microstrip line to be 54.5 mm which is quite close to the value obtained from equation $l = \frac{\lambda}{2\pi}\tan^{-1}\left(\frac{X_c}{Z_o}\right)$. The resistive part of the input impedance is 11.16 Ω and can be obtained by tuning the coupling distance d between the arc-shaped strip and the eccentric circular rings. Here, the coupling distance is chosen to bed=0.75 mm. Two shorting pins, of radius0.5 mm each, connect the arc-shaped feeding strip to the ground plane (of area 91×91 mm²) on the opposite side of the substrate.

Initially, only the larger ring (Antenna 1) of outer radius, r_1=59.9 mm, and inner radius, r_2= 46 mm, is taken and excited by the arc-shaped strip. It is found to resonate at 866 MHz with a reflection coefficient of−33.43 dB. Next, only the smaller ring (Antenna2) of outer and inner radii, r_3=26.6mm andr4=23.4 mm, respectively, is considered with the feed structure. The inner ring resonates at 956 MHz. The value of the reflection coefficient at 956 MHz is−21.13 dB. Finally, by placing the smaller ring (outer radius, R_3=31.9mmand inner radius, R_4= 24 mm) inside the larger one (outer radius, R_1=39.4 mm, and inner radius, R_2=35.4 mm), dual-band operation is obtained at 866 MHz and 956 MHz, with each of the rings working as a single band resonator. The simulated return loss of the mono-band antennas and that of the proposed antenna are shown in figure.

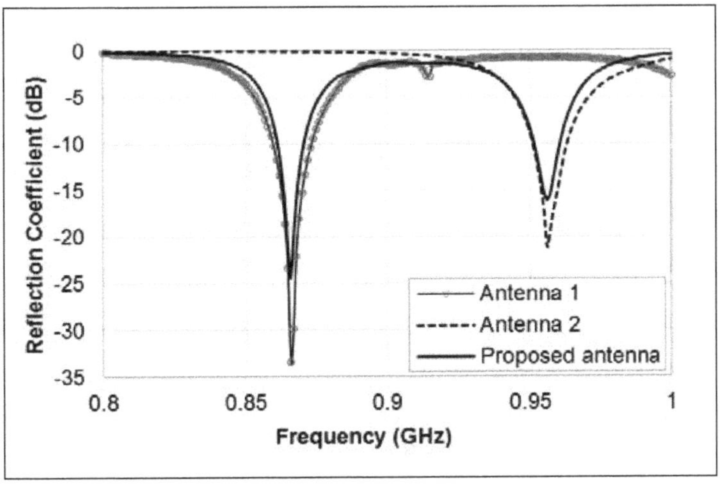

Reflection coefficients of Antenna 1, Antenna 2 and the proposed antenna.

The surface current distributions of the proposed tag antenna at the higher and lower resonant frequencies are illustrated in figure. From figure, it can be clearly observed that at lower resonant frequency, i.e., at 866 MHz, the magnitude of current in the larger ring is more than that of the smaller ring, whereas the current distribution is more in the smaller ring than in the larger ring at the higher resonating frequency, i.e., at 956 MHz. Hence, it can be concluded that the smaller ring is responsible for resonance at the upper frequency band, and the larger ring is responsible for resonance at the lower frequency band.

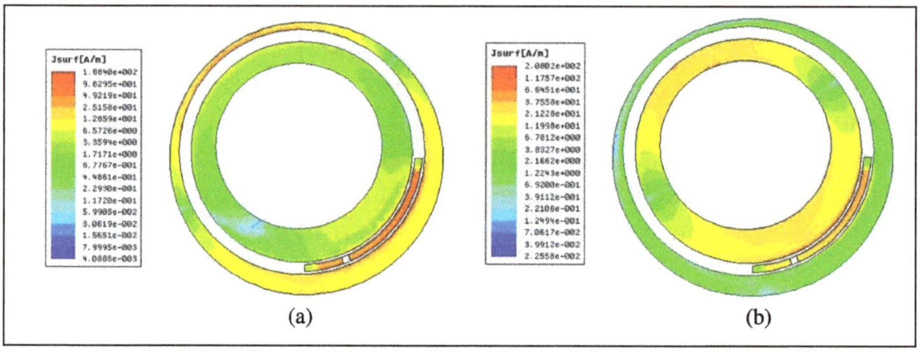

Simulated surface current distribution on the eccentric circular rings at (a) 866 MHz and (b) 956 MHz.

Parametric Analysis

The Effects of Varying R_1 and R_4

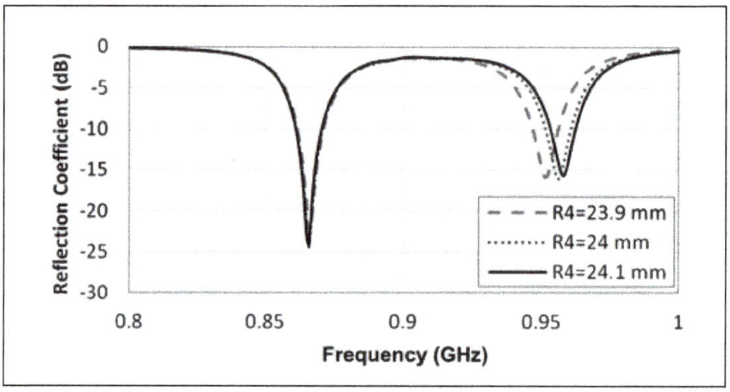

Effects of varying R_1 from 39.3 mm to 39.5 mm on the reflection coefficient.

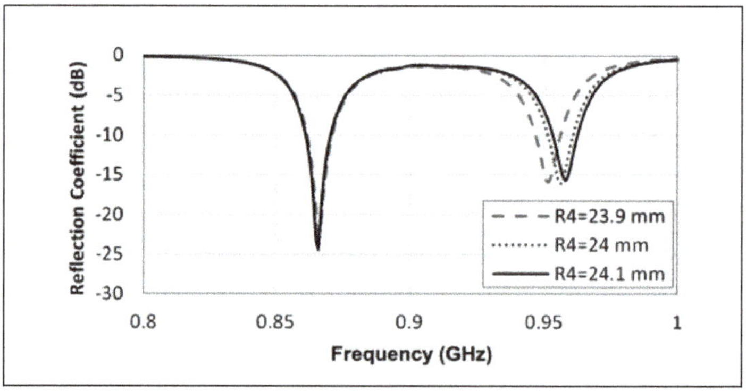

Effects of varying R_4 form 23.9 mm to 24.1 mm on the reflection coefficient.

In the proposed dual-band tag antenna, the outer ring is responsible for the resonance at the lower band and the inner ring for the higher frequency band. If R_1 is increased, keeping all other parameters constant, the dimension of the larger ring increases. This decreases the resonant frequency of the lower band. Figure shows that with varying R_1 from 39.3 mm to 39.5 mm, the resonant frequency shifts from 869 MHz to 864 MHz without affecting the higher band. Similarly, by changing the value of R_4, keeping other parameters unchanged, the higher band can be tuned without much influence on the lower band. The effects of varying R_4 are shown in figure. As R_4 is increased from 23.9 mm to 24.1 mm, the dimension of the smaller ring decreases, which in turn shifts the upper band resonating frequency from 952 MHz to 959 MHz.

The Effects of Tuning β

The chip is attached to the short-circuited microstrip feed-line making an angle β with positive x-axis. As β is decreased, both the real and imaginary parts of impedance increase. Thus, it is observed that with varying β proper impedance matching is obtained. Figure below shows different positions of the chip as β is decreased from 45° to 22°, and their corresponding impedances are plotted. The best impedance matching is obtained at β=22° where the simulated impedances of the antenna are found to be 11.7+j151.7Ω at 866 MHz and 7.3+j136.7Ω at 956 MHz resulting in return losses of−24.4d Band−16.1 dB, respectively.

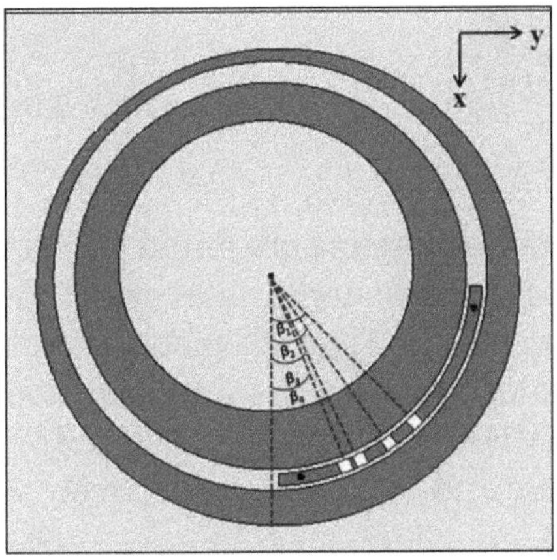

The proposed antenna showing different positions of the chip as βis varied from 45° to 22°.

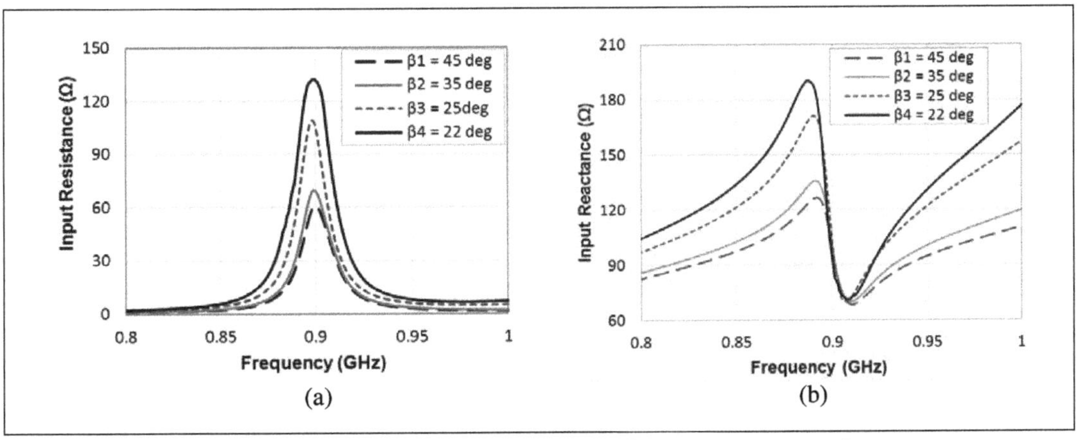

Variation in the (a) input resistance (b) input reactance of the proposed antenna on decreasing β from 45° to 22°.

Reflection Coefficient

The proposed dual-band tag antenna is fabricated by using the standard etching technique with PCB Prototyping machine (LPKF ProtoMat S103). Figure below shows the fabricated antenna. Of the different input impedance measurement techniques that can be employed for RFID tag antennas (using baluns probes, mirror theories, differential probe), the differential probe method is used hereto measure the input impedance at both the frequencies. Anritsu vector network analyzer (Model No.MS2038C) is used for the measurement of the input impedance through differential probes. It shows the simulated and measured impedance plots of the antenna.

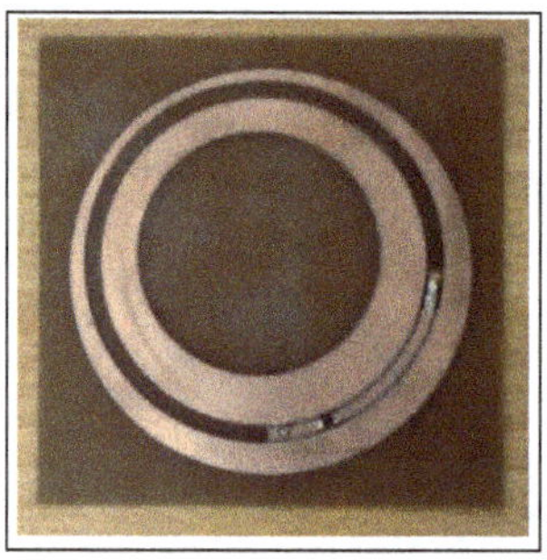

Fabricated antenna.

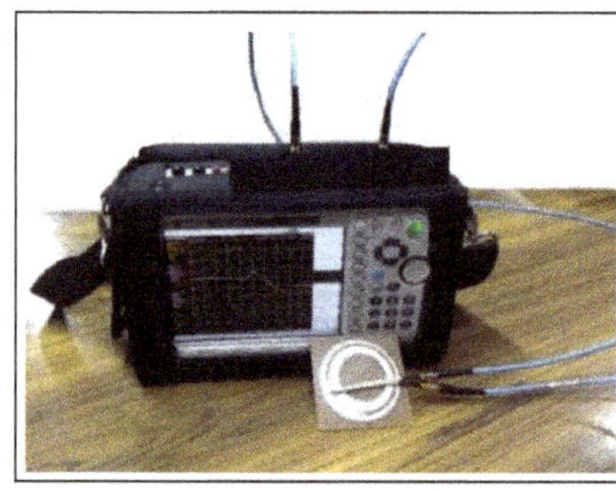

Measurement setup using VNA.

The simulated impedances of the antenna are found to be 11.7+j151.7Ω at 866 MHz and 7.3+j136.7Ωat 956 MHz, and the measured values are 21.2+j149.2Ω and 14 +j143.6Ω at 866 MHz and 956 MHz, respectively. Figure depicts the simulated and measured reflection coefficients. The simulated values of the reflection coefficient are−24.4 dB at 866 MHz and−16.10 dB at 956 MHz, whereas the measured values of the same are−12.25 dB and−12.99 dB at 866 MHz and 952 MHz, respectively.

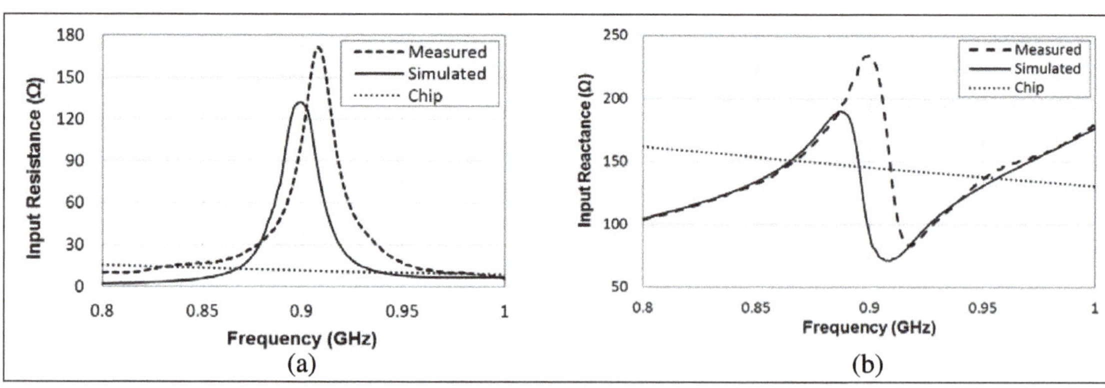

Measured and simulated (a) input resistance (b) input reactance of the proposed antenna.

The simulated and measured results are found in good agreement with each other. It is observed that the measured 10 dB bandwidth is 9 MHz from 862 to 870 MHz and 8 MHz in the range of 949–956 MHz covering the UHF RFID bands used in Europe and Japan.

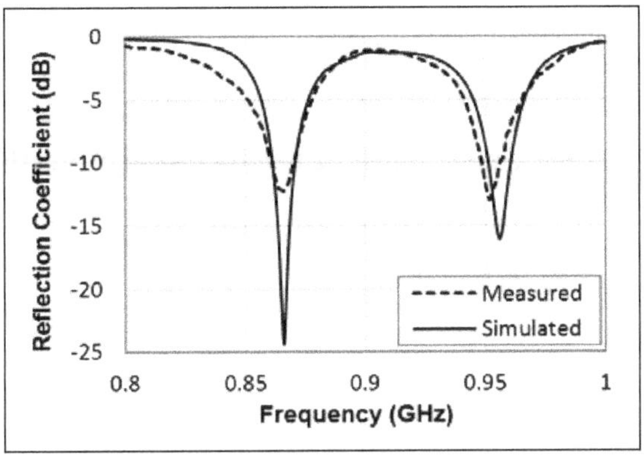

Simulated and measured reflection coefficients of the proposed antenna.

Read Range

To calculate the read range of the tag, Friis' free-space formula is used, as given below:

$$r = \frac{\lambda}{4\pi}\sqrt{\frac{P_t G_t G_r \tau}{P_{th}}}$$

Where λ is the wave length; P_t is the power transmitted by the reader; G_t and G_r are the gains of the transmitting and receiving antennas, respectively; P_{th} is the minimum threshold power required to activate the RFID tag IC chip (for Impinj Monza 4 IC chip, P_{th}=−17.4dBm); τ is the power transmission coefficient which is given by,

$$\tau = \frac{4R_{chip}R_a}{\left|Z_a + Z_{chip}\right|^2}, \quad 0 \le \tau \le 1$$

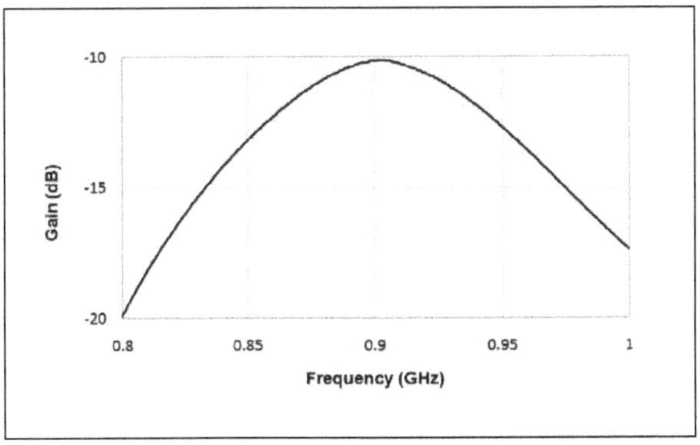

Simulated gain versus frequency plot of the proposed antenna.

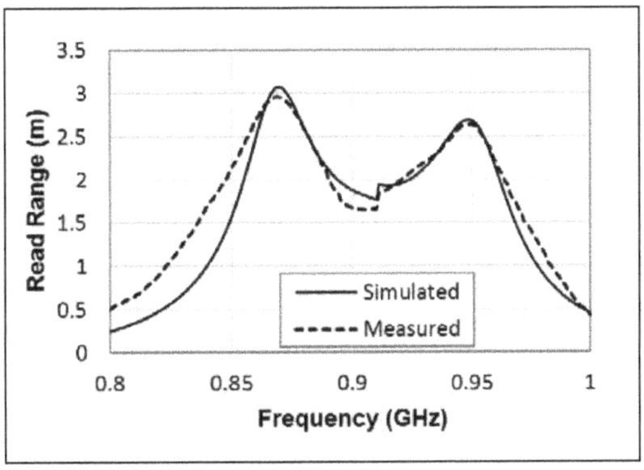

Simulated and measured read ranges.

Where $Z_{chip} = R_{chip} + jX_{chip}$ the impedance of the IC is chip, and $Z_a = R_a + jX_a$ is the impedance of the tag antenna. The product of the power transmitted by the reader and the gain of the transmitting antenna (i.e., P_tG_t) is the transmitter's effective isotropically radiated power (EIRP), which is allotted for different regions as per the International

Telecommunication Union frequency regulations. The value of EIRP in European frequencies is 3.3 W, whereas in Japan it is 4 W. The simulated gain of the antenna at different values of frequency is plotted in figure below. Using these values of gain, the read range is calculated at each frequency by employing equation. The plots of the measured and theoretical read ranges of the proposed antenna are shown in figure. The theoretical and measured read ranges in the 865–868 MHz band and 950–956 MHz band are found around 3 m and 2.6 m, respectively.

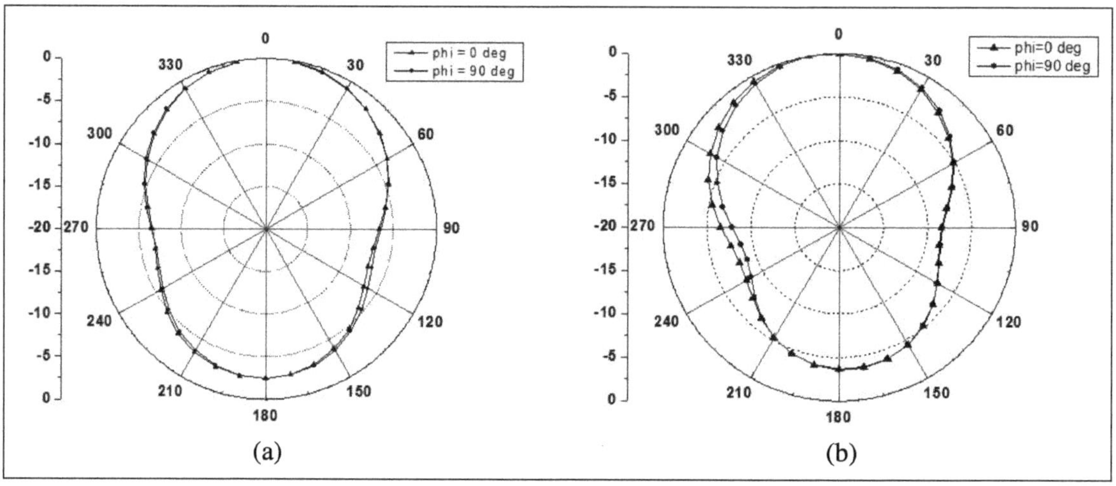

Radiation patterns at (a) 866 MHz (b) 956 MHz.

Radiation Pattern

The radiation patterns of the antenna at 866 MHz and 956 MHz are simulated and found bidirectional, which are desirable for RFID applications. The simulated radiation patterns at both the frequencies.

A comparison of the proposed antenna with some of the UHF dual-band RFID tag antennas reported in the literature is shown in table below. Although the antennas reported in are smaller in dimension than the proposed antenna, considerable increment in the read range has been achieved in the designed antenna.

Table: Comparison of the proposed antenna with the previously reported UHF dual-band RFID tag antennas.

Antenna Size	Operating Bands	Read Range (m)
20×33.5	865–868 MHz	0.75
	908.5–914 MHz	1.6
65×52	860 MHz	2.5
	920 MHz	2
91×91	865–868 MHz	3
	950–956 MHz	2.6

ACTIVE TAGS

Active RFID (radio frequency identification) tags are continuously operating, battery-powered sensors that gather and transmit data to a reading device.

An active RFID system consists of a reader, tag and antenna. Unlike passive RFID tags that contains merely an antenna and a microchip with no internal power source, an active RFID tag has its own power source -- an on-board, long-lasting battery that enables the tag to transmit data continuously, regardless of whether it's in the field range of a reader.

There are two kinds of active RFID tags: transponders and beacons. A transponder only communicates when it's in the immediate vicinity of a reader. A beacon broadcasts constantly. A number of unique characteristics are specific to active RFID tags. Because the tags are often required to survive in harsh environmental conditions like extreme temperatures or precipitation, some are encased in a protective shell. Given the size requirements of the battery, wiring and the durable exterior, active RFID tags are larger than passive tags. Some also have on-board sensors that track environmental parameters like the temperature, humidity, fluid and power consumption. They can also be used for highly specific use cases, such as sensing when a container door is open and there is a change in temperature. This gives shippers an extra insight that allows them to make important adjustments, like rerouting a shipment if it spoils.

Active RFID tags require low signal strength to communicate and can broadcast up to and even beyond a range of 100 meters. The cost per tag ranges anywhere from $15 to more than $100, depending on its capabilities. The high cost generally makes active RFID tags too expensive for simple inventory applications. They are better for tracking high-value items like cargo. Industries that commonly use active RFID tags in addition to shipping and logistics include automobile sales manufacturing, health and medical, construction, mining, remote monitoring and IT asset management.

ANTENNA

An RFID reader communicates with a tag through the reader antenna which broadcasts the RF signals from a reader transmitter to its surroundings and receives the response from the tags. Antenna is one of the key factors for RFID systems; the detection range and accuracy are directly dependent on the performance of reader / tag antennas. In addition, optimized antenna design will benefit the RFID system with longer reading range, better accuracy, reduced antenna fabrication cost, and simple system configuration and implementation.

Since the frequency available for RFID applications varies within a broad frequency range from LF (125–134.2 kHz), high frequency (HF) (13.56 MHz), ultra-high frequency

(UHF) (840–960 MHz) to MWF (2.4 GHz, 5.8 GHz, and 24 GHz), the RFID antenna designs feature distinct requirements and variations. Therefore, some unique issues should be taken into account when designing RFID antennas.

Loop antennas have been widely used in LF / HF RFID systems as reader antennas for many years since most of the LF/HF RFID systems are based on inductively coupling for reader / tag power transferring and communications. The main design considerations of these electrically small loop antennas include coupling zone size, magnetic field strength, quality factor, impedance matching, environment, and so on.

When the operating frequencies rise to UHF band, the conventional solid-line loop antenna cannot operate properly since the perimeter of the loop antenna may be of several operating wavelengths for adequate coverage. The electrically large solid-line loop antenna cannot generate even magnetic field distribution in the near-field zone of the antenna because the current distribution along the loop experiences phase-inversion and current nulls, which degrades the reliability of RFID systems. The major challenge to design such UHF near-field RFID reader antennas is to ensure electrically large loop antennas producing the strong and uniform magnetic field distribution over a larger interrogation zone.

Far-field Reader Antenna

Far-field RFID reader antennas have a number of variations wherein the most commonly used is the circularly polarized antenna for detecting arbitrarily oriented tags. Linear antennas such as dipole antennas have also been adopted in some specific applications where the orientation of the tags is fixed. Generally, the major considerations in far-field reader antennas include frequency range / bandwidth, polarization, gain, axial ratio, impedance matching, size, cost, and mechanical robustness. A broadband antenna which features desirable performance across the entire UHF RFID band of 840–960 MHz (13.3%) would be preferable for system configuration, implementation, and cost reduction.

RFID Tag Antenna

A typical passive RFID tag consists of an antenna and a microchip. The characteristics of the microchip are quantified by IC manufacturers and cannot be modified by users. The key challenge to design a tag antenna is conducted with a prior selected microchip under the various constraints to achieve a maximum reading range. Generally, the following issues should be addressed when designing a tag antenna:

- Receiving maximum signals from a reader antenna to power up a microchip. The far-field tag antenna is required to be conjugate matched to the microchip. For near-field tag antenna (coil), proper inductance is required to configure a resonant circuit with chip capacitance at the operating frequency.

- Small/thin enough for being attached to or embedded into the specified object.
- Insensitive to the attached object to keep performance consistent.
- Required radiation patterns (omni-directional, directional or hemispherical).
- Robust in mechanical structure.
- Low cost in both materials and process.

In general, loop type antennas are most commonly used near-field RFID systems to capture the energy from the reader by magnetic field coupling. For far-field applications, various types of antennas such as dipole antenna, meander line antenna, slot antenna, and patch antenna are widely being employed.

CLIPPED TAG

The clipped tag is a radio frequency identification (RFID) tag designed to enhance consumer privacy. RFID is an identification technology in which information stored in semiconductor chips contained in RFID tags is communicated by means of radio waves to RFID readers. The most simple passive RFID tags do not have batteries or transmitters. They get their energy from the field of the reader. They transfer their information to the reader by modulating the signal that is reflected back to the reader by the tag. Because tags depend on the reader for power their range is limited, typically up to 10 meters or 30 feet for UHF RFID tags.

Today, the public uses RFID tags for many applications including electronic toll collection, E-ZPass for example, or the Speedpass which is used as a credit token for the purchase of gasoline. The retail supply chain uses RFID tags to monitor the passage of pallets and cases at loading dock doors. The expectation for the future is for RFID tags to be used for the labelling of items for retail sale. Concerns for individual privacy have been raised because the RFID tags may be read by invisible radio waves without the knowledge of the holder of the tagged item.

The privacy-protecting RFID tag, the "clipped tag" has been suggested by IBM. The clipped tag puts the option of privacy protection in the hands of the consumer. After the point of sale, a consumer may tear off a portion of the tag, much like the way in which a ketchup packet is opened. This allows the transformation of a long-range tag into a proximity tag that still may be read, but only at short range – less than a few inches or centimeters. The modification of the tag may be confirmed visually. The tag may still be used later for returns, recalls, or recycling. The clipped tag was listed among the Wall Street Journal Technology Innovation Winners for 2006. Two US patents were issued for this invention in 2007.

Other mechanisms designed to protect privacy for RFID item tagging for retail use are the EPCglobal kill command and the RSA blocker tag.

EAR TAG

A sheep with an ear tag.

An ear tag is a plastic or metal object used for identification of domestic livestock and other animals. If the ear tag uses Radio Frequency Identification Device (RFID) technology it is referred to as an electronic ear tag. Electronic ear tags conform to international standards ISO 11784 and ISO 11785 working at 134.2 kHz, as well as ISO/IEC 18000-6C operating in the UHF spectrum. There are other non-standard systems such as Destron working at 125 kHz. Although there are many shapes of ear tags, the main types in current use are as follows:

- Flag-shaped ear tag: Two discs joined through the ear, one or both bearing a wide, flat plastic surface on which identification details are written or printed in large, easily legible script.
- Button-shaped ear tag: Two discs joined through the ear.
- Plastic clip ear tag: A moulded plastic strip, folded over the edge of the ear and joined through it.
- Metal ear tag: An aluminium, steel or brass rectangle with sharp points, clipped over the edge of the ear, with the identification stamped into it.
- Electronic Identification Tags, include the EID number and sometimes a management number on the button that appears on the back of the ear. These can at times be combined as a matched set, which includes Visual tags with Electronic Identification Tags.

Each of these except the metal type may carry a RFID chip, which normally carries an electronic version of the same identification number.

An ear tag usually carries an Animal Identification Number (AIN) or code for the animal, or for its herd or flock. Non electronic ear tags may be simply handwritten for the convenience of the farmer (these are known as "management tags"). Alternatively this identification number (ID) may be assigned by an organisation, such as the Meat and Livestock Association (MLA), which is a not-for-profit organisation owned by cattle, sheep and goat producers; funded by a levy on livestock sales with Federal Government input. Electronic tags may also show other information about the animal, including other related identification numbers; such as the Property Identification Code (PIC) for the properties the animals have been located. In the case of MLA's NLIS the movement of certain species of livestock (primarily cattle, goats and sheep) must be recorded in the online database within 24 hours of the movement; and include the PICs of the properties the animals are travelling between. The National Livestock Identification System (NLIS) of Australia regulations require that all cattle be fitted with a RFID device in the form of an ear tag or rumen bolus (a cylindrical object placed in the rumen) before movement from the property and that the movement be reported to the NLIS. However, if animals are tagged for internal purposes in a herd or farm, IDs need not be unique in larger scales. The NLIS now also requires sheep and goats to use an ear tag that has the Property Identification Code inscribed on it. These ear tags and boluses are complemented by transport documents supplied by vendors that are used for identification and tracking. A similar system is used for cattle in the European Union (EU), each bovine animal having a passport document and tag in each ear carrying the same number. Sheep and goats in the EU have a tag in one or both ears carrying the official number of their flock and also for breeding stock an individual number for each animal; one of these tags (usually the left) must have a RFID chip (or the chip may instead be carried in a rumen bolus or on an anklet).

An ear tag can be applied with an ear tag applicator, however there are also specially-designed tags that can be applied by hand. Depending on the purpose of the tagging, an animal may be tagged on one ear or both. There may be requirements for the placement of ear tags, and care must be taken to ensure they are not placed too close to the edge of the ear pinnae; which may leave the tag vulnerable to being ripped out accidentally. If there exists a national animal identification programme in a country, animals may be tagged on both ears for the sake of increased security and effectiveness, or as a legal requirement. If animals are tagged for private purposes, usually one ear is tagged. Australian sheep and goats are required to have visually readable ear tags printed with a Property Identification Code (PIC). They are complemented by movement documents supplied by consignors that are used for identification and tracking.

Very small ear tags are available for laboratory animals such as mice and rats. They are usually sold with a device that pierces the animal's ear and installs the tag at the same time. Lab animals can also be identified by other methods such as ear punching or marking, implanted RFID tags (mice are too small to wear an ear tag containing an RFID chip), and dye.

A sow polar bear with ear tag.

A tagged black-tailed prairie dog.

Livestock ear tags were developed in 1799 under the direction of Sir Joseph Banks, President of the Royal Society, for identification of Merino sheep in the flock established for King George III. Matthew Boulton designed and produced the first batch of sheep eartags, and produced subsequent batches, modified according to suggestions received from Banks. The first tags were made of tin.

Ear tags were incorporated as breed identification in the United States with the forming of the International Ohio Improved Chester Association as early as 1895, and stipulated in the topic of Incorporation, as an association animal and breed identification, of the improved Chester White.

Although ear tags were developed in Canada as early as 1913 as a means to identify cattle when testing for tuberculosis, the significant increase of use of ear tags appeared with the outbreak of BSE in UK. Today, ear tags in a variety of designs are used throughout the world on many species of animal to ensure traceability, to help prevent theft and to control disease outbreaks.

The first ear tags were primarily steel with nickel plating. After World War II, larger, flag-like, plastic tags were developed in the United States. Designed to be visible from a distance, these were applied by cutting a slit in the ear and slipping the arrow-shaped head of the tag through it so that the flag would hang from the ear.

In 1953, the first two-piece, self-piercing plastic ear tag was developed and patented. This tag, which combined the easy application of metal tags with the visibility and colour options of plastic tags, also limited the transfer of blood-borne diseases between animals during the application process.

Some cattle ear tags contain chemicals to repel insects, such as buffalo flies, horseflies, etc. Metal ear tags are used to identify the date of regulation shearing of stud and show sheep. Today, a large number of manufacturers are in competition for the identification of world livestock population .

In 2004, the U.S. Government asked farmers to use EID or Electronic Identification ear tags on all their cattle. This request was part of the National Animal Identification System (NAIS) spurred by the discovery of the first case of mad cow disease in the United States. Due to poor performance and concern that other people could access their confidential information, only about 30 percent of cattle producers in the United States tried using EID tags using standards based on the low frequency standards, while the UHF standards are being mandated for use in Brazil, Paraguay, and Korea . The United States Department of Agriculture maintains a list of manufacturers approved to sell ear tags in the USA.

Ear tags (conventional and electronic) are used in the EU as official ID system for cattle, sheep and goat, in some cases combined with RFID devices.

The International Committee for Animal Recording (ICAR) controls the issue electronic tag numbers under ISO regulation 11784.

The National Livestock Identification System (NLIS) is Australia's system for tracing cattle, sheep and goats from birth to slaughter.

In Canada, the Health of Animals Regulations require approved ear tags on all bison, cattle and sheep that leave the farm of origin, except that a bovine may be moved, without a tag, from the farm of origin to a tagging site. RFID (radio frequency identification) tags are used for cattle in Canada and metal as well as RFID tags have been in use for sheep. Mandatory RFID tagging of sheep in Canada will be deferred to some later date.

Other Forms of Animal Identification

Pigs, cattle and sheep are frequently earmarked with pliers that notch registered owner and age marks into the ear. Mares on large horse breeding farms have a plastic tag attached to a neck strap for identification; which preserves their ears free of notches. Dairy cows are sometimes identified with ratchet fastened plastic anklets fitted on the pastern for ready inspection during milking; however NLIS requirements apply to cattle - including both dairy and beef animals. More commonly coloured electrical tape is used as short term ankle identifiers for dairy animals to identify when one teat should not be milked for any reason. Laboratory rodents are often marked with ear tags, ear notches or implantable microchips.

BAG TAG

Bag tags, also known as baggage tags, baggage checks or luggage tickets, have traditionally been used by bus, train, and airline carriers to route checked luggage to its final destination. The passenger stub is typically handed to the passenger or attached to the ticket envelope:

- To aid the passenger in identifying their bag among similar bags at the destination baggage carousel.
- As proof—still requested at a few airports—that the passenger is not removing someone else's bag from the baggage reclaim hall.
- As a means for the passenger and carrier to identify and trace a specific bag that has gone astray and was not delivered at the destination. The carriers' liability is restricted to published tariffs and international agreements.

Example of IATA airport code printed on a baggage tag, showing DCA.

The first automated baggage sorting systems were developed in the 1980s by Eastern Air Lines at their Miami International Airport hub. Other airlines soon followed with their own systems, including United Air Lines, TWA, Delta, and American Airlines. None of these systems were interchangeable. In some systems, the bar code was used to represent a three-letter destination airport code, and in others it was a two-digit sorting symbol instructing the system at which pier to deliver the bag.

As a result of the bombing of Air India Flight 182 on June 23, 1985, the airline industry, led by IATA, convened the Baggage Security Working Group (BSWG) to change international standards and require passenger baggage reconciliation. The Chairman of the BSWG, John Vermilye of Eastern Airlines, proposed that the industry adopt the already-proven license plate system.

This concept used a barcode to represent the baggage tag number. At check-in, this number was associated with the passenger details, including flight number, destination, connection information, and even class of service to indicate priority handling.

Working with Allen Davidson of Litton Industries, with whom Eastern had developed the license plate concept, the BSWG adopted this system as the common industry standard for passenger baggage reconciliation. Initially the barcode, or license plate, was used to match baggage with passengers, ensuring that only the baggage of passengers who had actually boarded the flight were carried onto the aircraft. This standard was adopted by IATA Resolution in 1987.

By 1989, the license plate concept was expanded to become the industry standard for automated baggage sorting as well. The barcodes were enlarged to facilitate automated reading. The barcode was shown in two different orientations or in a "T" shape, called the "orthogonal" representation.

The term *license plate* is the official term used by the IATA, the airlines, and the airports for the ten-digit numeric code on a bag tag issued by a carrier or handling agent at check-in. The license plate is printed on the carrier tag in barcode form and in human-readable form.

The license plate is the index number linking the Baggage Source Message (BSM), sent by a carrier's departure control system, to the airport's baggage handling system. This message (BSM) contains the flight details and passenger information. Each digit in the license plate has a specific meaning. The automated baggage handling system scans the barcodes on the carrier tags and sorts the bags accordingly. Both the license plate number and the BSM are essential for automated sorting of baggage.

The human-readable license plate will have either a two-character or a three-digit IATA carrier code. For example, it may be either "BA728359" or "0125728359." "BA" would be the two-character IATA code for British Airways), and "125" would be the three-digit IATA carrier code. Nevertheless, the barcode will always be the full ten digits.

The first digit in the ten-digit license plate is not part of the carrier code. It can be in the range of zero to nine. Zero is for interline or online tags, one is for fallback tags, and two is for "rush" tags.

Fallback tags are pre-printed or demand-printed tags for use only by the airport's baggage handling system. These tags are used when there is a problem in communication between the carrier's departure control system and the airport's baggage handling system. A "rush" bag is a bag that missed its original flight and is now flying unaccompanied.

The purpose of numbers in the range of three to nine is not defined by the IATA, but they can be used by each carrier for their own specific needs. The first digit is commonly used as a million indicator for the normal six-digit tag number.

Besides the license plate number, the tag also has:

- Name of Airport of Arrival.
- Departure Time.
- IATA airport code of Airport of Arrival.

- Airline Code and Flight Number.
- Name of Passenger Identified with the Baggage (Last Name, First Name).

ORGANIC RFID TAGS

The organic RFID tag consists of 4 different modules: the antenna coil, the HF-capacitor, the rectifier and the transponder chip with an integrated load modulator. The coil and the HF-capacitor form an LC tank resonating at the HF resonance frequency of 13.56 MHz, which energizes the organic rectifier with an AC voltage at 13.56 MHz. The rectifier generates the DC supply voltage for the 64-bit organic transponder chip, which drives the modulation transistor between the on- and off-state with a 64-bit code sequence. Load modulation can be obtained in two different modes, depending on the position of the load modulation transistor in the RFID circuit. AC load modulation, whereby the modulation transistor is placed in front of the rectifier, sets demanding requirements to the OTFT, since it has to be able to operate at HF frequency. This is not obvious, as a consequence of the limited charge carrier mobility of the OTFT, 0.1 – 1 cm2/Vs for pentacene as organic semiconductor. Therefore, load modulation at the output of the rectifier (DC load modulation) is preferred in organic RFID tags. In latter mode, the OTFT does not require to operate at HF frequency. The organic RFID tags operate in DC load modulation mode. Nevertheless, organic RFID tags operating in AC load modulation mode have also been achieved.

Technology

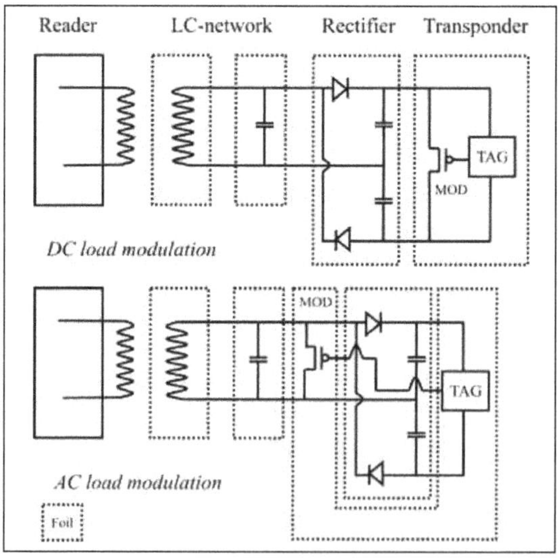

Inductively-coupled organic RFID tags using DC (top pane) and AC (bottom panel) load modulation.

In this part we describe the technology used to create high-performance organic RFID tags. As mentioned earlier, the tags are composed of four flexible foils, with the following components: an inductor coil, a capacitor, a rectifier and a transponder. The coil is made from etched copper on foil, and was manufactured by Hueck Folien GmbH. The HF-capacitor consists of a metal-insulator-metal stack (MIM stack), processed on a 200- μm thick flexible polyethylene naphthalate (PEN) foil (Teonex Q65A, Dupont Teijin Films). The insulator material used for the capacitor is Parylene diX SR.

The rectifier comprises two vertical Schottky diodes, and two capacitors in a so-called double half-wave configuration. The substrate for manufacturing the rectifiers is a 200 μm thick, flexible 150 mm PEN foil, on which first a metal-insulator-metal (MIM) stack is processed for the capacitors in the circuit. The metal layers are 30 nm of gold (Au) and the insulator is Parylene diX SR, with a relative dielectric constant of εr of 3 and a thickness of 400 nm. Conventional photolithography is used to define the capacitors in the MIM stack. The top Au layer of the MIM stack is used as anode for the vertical diode. A 350 nm pentacene layer, the organic semiconductor, is evaporated through a shadowmask by HV-deposition. Last, an aluminum (Al) cathode is evaporated through a second shadowmask.

The organic 64-bit transponder chip is made on a 25 μm thin plastic substrate using organic bottom-gate thin-film transistors. The organic electronics technology that is used, was developed by Polymer Vision for commercialization in rollable active matrix displays and is described elsewhere. The insulator layers and the semiconductor layer are organic materials processed from solution. The transistors, with a typical channel length of 5 μm, exhibit an average saturation mobility of 0.15 cm2/Vs. A micrograph picture of the 64-bit transponder chip and the 6" wafer is depicted in figure.

The RFID Measurement Setup

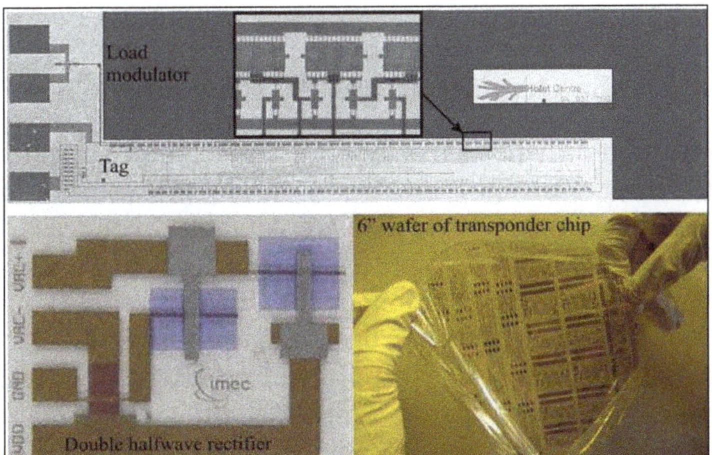

Pictures of the transponder foil and the load modulator foil (top), the double halfwave rectifier foil (bottom left) and the 6" wafer full of transponder chips (bottom right).

The complete tag is realized by properly interconnecting the contacts of the four foils, which we achieved in an experimental set-up where we plug the individual foils into sockets as shown in figure. Alternatively, we have also achieved tags by lamination of the foils, whereby electrically conductive glue is used to interconnect the different contacts of the individual foils.

The reader setup conforms to the ECMA-356 standard for "RF Interface Test Methods". It comprises a field generating antenna and two parallel sense coils, which are matched to cancel the emitted field. By this method, only the signal sent by the RFID tag is read out at the reader side. The detected signal is then demodulated by a simple envelope detector, being a diode followed by a capacitor and a resistor, and shown on an oscilloscope.

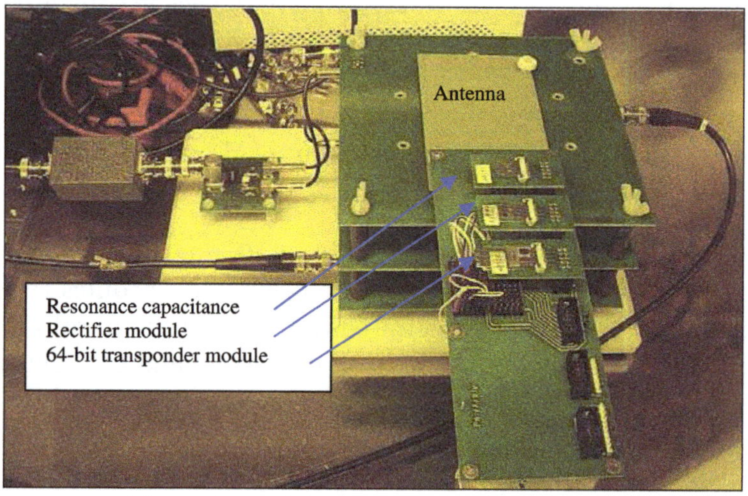

The reader and RFID tag measurement setup.
Foils are placed in sockets to ease manipulation.

Organic Transponder Chip

Design Issues for the Organic RFID Transponder Chip

In this part, we will discuss challenges in designing large, robust circuitry in present-day OTFT technology. A first challenge is created by the fact that in common OTFT technologies only unipolar, organic semiconductors are available. In our case, the design of the organic RFID transponder chip was limited to p-type only logic, since pentacene is a hole transport semiconductor. In addition, only a single threshold voltage is present in this technology. As a consequence of this, designs based on complementary logic or dual-threshold voltage logic (e.g. nMOS) cannot be implemented. Therefore, in the field of organic electronics, research towards complementary logic has gained considerable attention, potentially leading to an improvement of the robustness of organic circuitry. Very recently, the first 4-bit complementary organic RFID tag has been demonstrated.

A big challenge in designing complex organic circuitry is the intrinsic parameter spread, in particular on the threshold voltage and the mobility. To tackle this problem in the design, we used a Monte Carlo tool to distribute the transistor parameters randomly for each OTFT in the circuit and to vary these parameters randomly for every new simulation according to the nominal value and within the given spread values for each parameter. The outcome of multiple simulations provides an idea of the robustness of the design and allows to investigate design improvements for increased robustness. To allow such statistical design, the number of transistors in the organic RFID transponder chip has been kept very limited.

Possible gate design architectures in unipolar, single threshold voltage logic are depletionmode logic or enhancement-mode logic. The gates in our organic RFID transponder chip are designed using the former, which is also known as zerovgs-logic. The choice for this type of logic was driven by the fact that the OTFTs used in this work show normally-on or depletion-mode behavior. The final design of the organic transponder chip has been made using only inverters and NAND-gates, both implemented in the zerovgs-logic as depicted in figure. The gain of such an inverter at the trip point, for supply voltages of 10 and 20 V, is 1.75 and 2.25, respectively. Nineteen-stage ring oscillators of inverters operate at a frequency of 627 Hz using 10 V supply voltage and at 692 Hz using 20 V supply voltage.

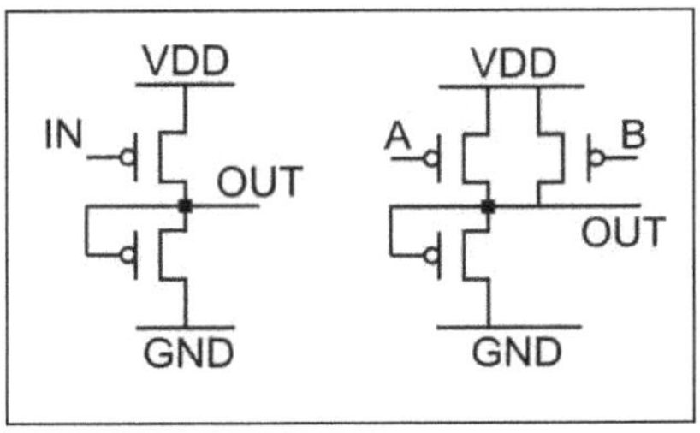

p-type only inverter (left) and NAND-gate (right) using the zerovgs-logic.

Final Design of the Transponder Chip

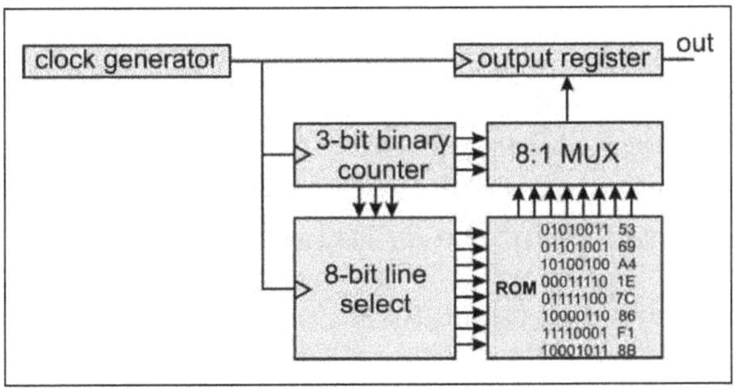

The digital logic part of the 64-bit transponder chip.

The schematic of the transponder foil is depicted in figure. A 19-stage ring oscillator generates the clock signal when powered. This clock signal is used to clock the output register, the 3-bit binary counter and the 8-bit line select. The 8-bit line select has an internal 3-bit binary counter and a 3-to-8 decoder. This block selects a row of 8 bits in the code. The 3-bit binary counter drives the 8:1 multiplexer, selecting a column of 8 bits in the code matrix. The data bit at the crossing of the active row and column, is transported via a 8:1 multiplexer to the output register, which sends this bit on the rising edge of the clock to the modulation transistor. The 3 bits of the 3-bit binary counter are also used in the 8-bit line select block for selecting a new row after all 8 bits in a row are transmitted.

The 64-bit transponder foil comprises only 414 OTFTs. Figure shows a micrograph image of the transponder foil. At 14 V supply voltage the 64-bit transponder foil generates the correct code at a data rate of 752 b/s, which is depicted in figure. Besides the 64-bit transponder foil, we also designed an 8-bit transponder foil. The main difference in the design is the complexity of the line select.

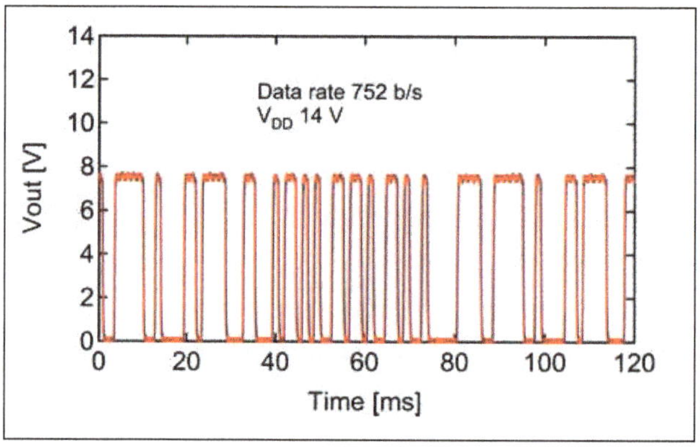

Measured code of the 64-bit transponder chip at a supply voltage of 14 V.

Organic Rectifier

Organic Diode

The purpose of the rectifier in an RFID tag is to create a DC-voltage from the AC-voltage detected and generated by an antenna at the targeted base carrier frequency of 13.56 MHz. This frequency is selected because it is a standard in Si-based RFID tags, and will therefore enable partial compatibility with installed reader systems at 13.56 MHz. An important issue for organic RFID tags is the efficiency of the rectifier. A more efficient rectification will result in the required DC-voltage from lower AC-input voltage. This implies larger reading distances for the RFID tags.

A rectifier comprises diodes and capacitors. For organic diodes, two different topologies can be used, being a vertical Schottky diode and a transistor with its gate shorted to its drain. The transistor with shorted gate-drain node is often considered as the most favorable topology because its process flow is equal to that used for the transistors in the digital circuit of the RFID tag. In this work, however, we have chosen to use the vertical diode structure because of its better intrinsic performance at higher frequencies compared to transistors as diodes.

The structure of a vertical, organic diode used to make the rectifier, is drawn in figure. As depicted, we fabricated hole-only organic diodes with a layer of pentacene (as organic semiconductor) sandwhiched between an Au- and Al-electrode on a 150 mm PEN foil. Because of their workfunctions, the Al-electrode blocks the injection of holes, whereas the Au-electrode permits the injection of holes.

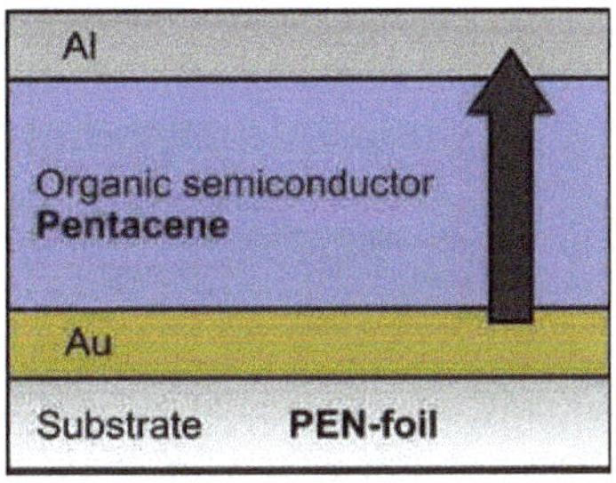

Structure of a vertical, organic Schottky diode.

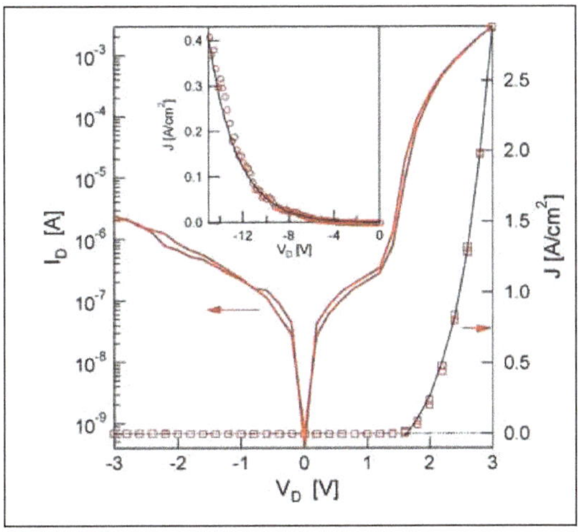

|I|-V and J-V characteristics of an organic pentacene diode (area=500x200 μm2) on linear (right axis) and logarithmic (left axis) scale. The inset shows the diode leakage current density between 0 and -15 V. The fits to the data of the current densities, both in forward and in reverse biases are represented by the solid black lines.

The electrical behavior of a single organic diode is examined by measurements performed in a nitrogen-filled glove box using an Agilent 4156C parameter analyzer. The current-voltage characteristics of a single pentacene diode are depicted in figure. The onset voltage of the diode is at about 1.2 V. However, 2 V is required to obtain a current density beyond 200 mA/cm². At 3 V forward bias, the current density is 2.88 A/cm². The leakage current of the diode shows an important increase when the reverse bias increases, limiting the maximum reverse voltage over the diode.

Organic Double Half-wave Rectifier

A double half-wave rectifier comprises two diodes, each followed by a capacitor. Figure shows the schematic of this circuit. Both capacitors are 20 pF. The active area of the diodes is 500 μm x 200 μm.

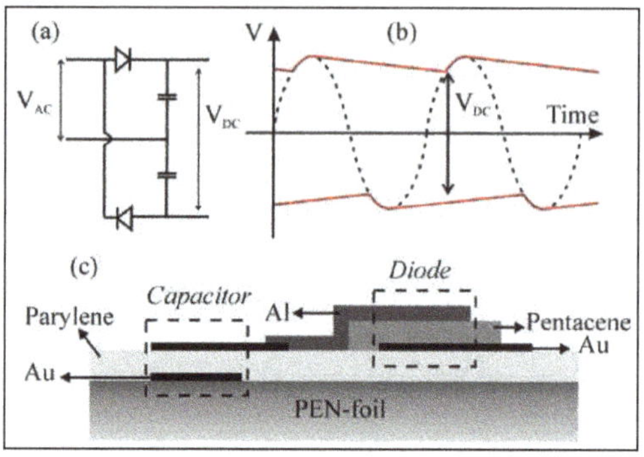

(a) Schematic of a double half-wave rectifier, (b) schematic operation of a double halfwave rectifier, (c) vertical cross-section of the integrated capacitors and diodes on foil.

A double half-wave rectifier circuit consists of two single half-wave rectifiers connected between the same nodes, with diodes connected as shown in figure. Both single half-wave rectifiers rectify the AC input voltage: one rectifies the upper cycles of the AC input voltage, the other single half-wave rectifier rectifies the lower cycles of the input voltage. This is schematically depicted in figure. The power and the ground voltage for the digital logic of the RFID tag are taken between both rectified signals. Therefore, a double half-wave rectifier yields about double the rectified voltage compared to a single half-wave rectifier.

For AC measurements of the rectifiers, we constructed a small printed circuit board (PCB), comprising coax-connections for the AC input signal and the DC output signal. The AC signal is terminated with 50 Ω for limiting reflections at 13.56 MHz. The DC signal is sent to the oscilloscope (MSO6014A of Agilent Technologies) on which the DC voltage was read out, having an input impedance of 1 MΩ, which is the load of the rectifier. The rectifier foil is placed into a socket which is plugged in the PCB. We used a Si-based double half-wave rectifier circuit (assembly of 1N4148-diodes and 20 pF capacitors) as a reference. Figure plots the rectified DC voltage of the integrated organic and assembled Si-based rectifiers versus the AC input voltage at a frequency of 13.56 MHz. At an amplitude of the AC voltage of 10.9 V, the rectified DC voltage of the organic rectifier is 14.9 V, while the Si-based rectifier produces 20.6 V. The slope of the measured V_{DCout} as a function of applied V_{ACin}, which we term the efficiency of the rectifier, is 1.64 for the organic rectifiers and 2 for the silicon rectifier. The onset voltage of the diodes can be estimated by extrapolation to an output voltage of 0 V and is 1.98 V for the organic rectifier and 0.42 V for the silicon rectifier. Each organic diode, therefore, has an onset voltage of only about 1 V.

To get a better understanding of the efficiency limitations of the organic-based rectifier, in particular the slope efficiency of $V_{DCout}/V_{ACin} = 1.64$, we model the behavior of the circuit of figure using the characteristics of figure. Both the forward and the reverse current density J can be expressed quite accurately as a function of the voltage V using the exponential curve $J(V) = J_0 + J_1 \exp(\alpha V)$. In the relevant forward current range (1.6 V to 3 V), $J_0 = -0.311$ A/cm², $J_1 = 0.0189$ A/cm² and $\alpha = 1.71$ V^{-1}. In the relevant reverse current range (-15 V to -1 V), $J_0 = 0$ A/cm², $J_1 = 0.001$ A/cm² and $\alpha = -0.4$ V^{-1}. These fits, extracted from the quasi-static characteristics, have been used to calculate the output voltage (V_{DC}) as a function of the AC input amplitude (V_{in}) by numerically matching the charging and discharging current of the rectifier's capacitors using below equation, derived from the equations and conditions outlined in Steudel et al. with A being the device area (0.001 cm²) and R_L being the load resistance (1 MΩ). For the commercial silicon diode, the curves from the datasheet have been used. Figure compares the calculated output voltage with the measured voltage. The slope efficiency is almost exactly reproduced. We can now trace back the origin of the slope efficiency to the presence of the reverse leakage current of the organic diodes: at higher input voltages, the leakage current increases exponentially, which discharges the capacitors, resulting in a lower DC output voltage. On the other hand, the modeled curve overestimates the DC output voltage by approximately a constant amount of 1 V, or 0.5 V for each diode. This has not been elucidated at present, and could be attributed to several factors, for example a slightly higher offset voltage or a slightly degraded mobility of the charge carriers by bias stress during the prolonged measurement time. We conclude from this analysis that the efficiency of the rectifier depends on the on/off ratio of the diode current.

$$\frac{1}{2\pi}\int_{\pi+\sin^{-1}\left(\frac{V_{DC}}{2V_{in}}\right)}^{2\pi+\sin^{-1}\left(\frac{V_{DC}}{2V_{in}}\right)} J\left(V_{in}\sin(\theta)-\frac{V_{DC}}{2}\right)\partial\theta+\frac{V_{DC}}{2AR_L}=\frac{1}{2\pi}\int_{\sin^{-1}\left(\frac{V_{DC}}{2V_{in}}+\frac{1.6}{V_{in}}\right)}^{\pi-\sin^{-1}\left(\frac{V_{DC}}{2V_{in}}+\frac{1.5}{V_{in}}\right)} J\left(V_{in}\sin(\theta)-\frac{V_{DC}}{2}\right)\partial\theta$$

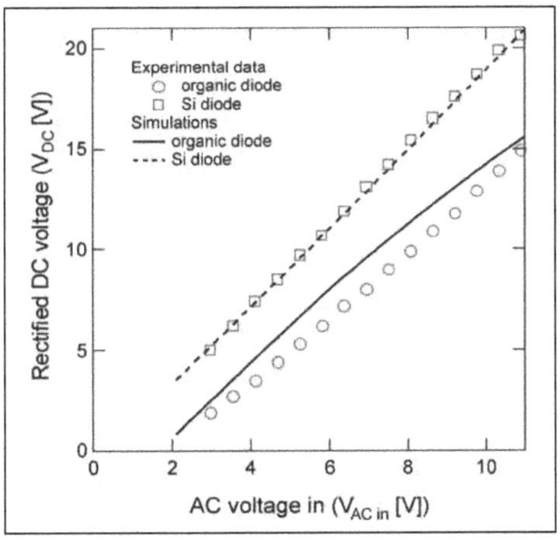

Simulation results and experimentally obtained data points of the rectified DC output voltage versus the AC input voltage for an organic-based rectifier and a Si-based rectifier both in double half-wave rectifier configuration.

Organic RFID Tag using DC load modulation

In this part, the organic RFID transponder foils are integrated with the antenna coil, the HFcapacitor and the rectifier to form an organic RFID tag. In DC load modulation mode, the modulation transistor (W/L = 5040μm/5μm) is placed behind the rectifier. All foils are placed into a socket and connected as depicted in figure. The RFID reader is a 7.5 cm radius antenna which emits the field at a base carrier frequency of 13.56 MHz.

In figure, the internal rectified voltage of this double half-wave rectifier in the organic RFID tag is plotted as a function of the field generated by the reader, for the tag antenna placed in the near-field of the reader antenna, at a distance of about 4 cm from the coil generating the readers 13.56 MHz RF field. As can be seen in the graph, 10 V rectified voltage is obtained at a 13.56 MHz electromagnetic field of about 0.9 A/m, and 14 V at 1.26 A/m.

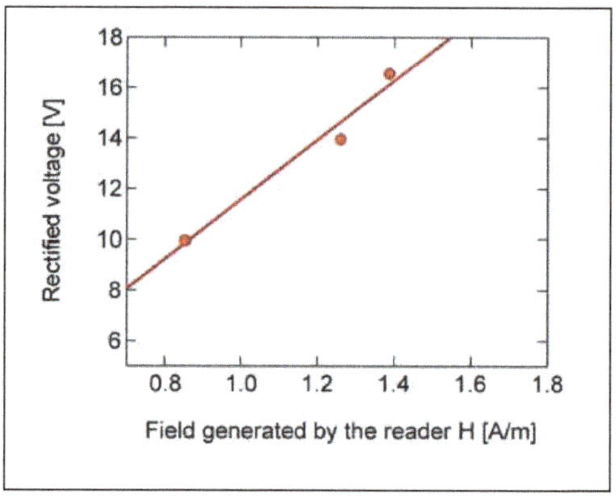

Internal rectified voltage of a double half-wave rectifier generated in an organic RFID tag versus the 13.56 MHz magnetic field generated by the reader.

The latter is the voltage currently required by our 64-bit organic transponder chips. The ISO 14443 standard states that RFID tags should be operational at a minimum required RF magnetic field strength of 1.5 A/m. The double half-wave rectifier circuit presented here therefore satisfies this ISO norm. After extrapolation of the measurement data, a DC voltage of 17.4 V can be obtained at a field of 1.5 A/m. If a single half-wave rectifier was used, the rectified voltage would be limited to 8 - 9 V, which is too low for current organic technology.

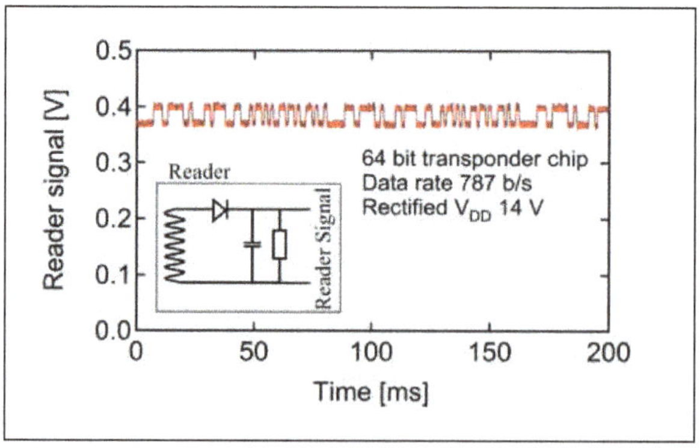

Signal of the 64-bit RFID tag measured on the reader (unamplified reader signal). The envelope detector of the reader is depicted in the inset.

The obtained rectified 14 V drives the transponder chip, which sends the code to the modulation transistor. The signal sent from the fully integrated, plastic tag is received by the reader and subsequently visualized using a simple envelope detector without amplification. The signal measured at the reader side is depicted in figure. This shows the fully functional, 64-bit RFID tag using an inductively-coupled 13.56 MHz RFID configuration with a data rate of 787 b/s. With a 0.7 V drop over the diode at the reader (envelope detector), a tag-generated signal of about 1.1 V is obtained, from which 30 mV is load modulation (modulation depth h = 1.4%).

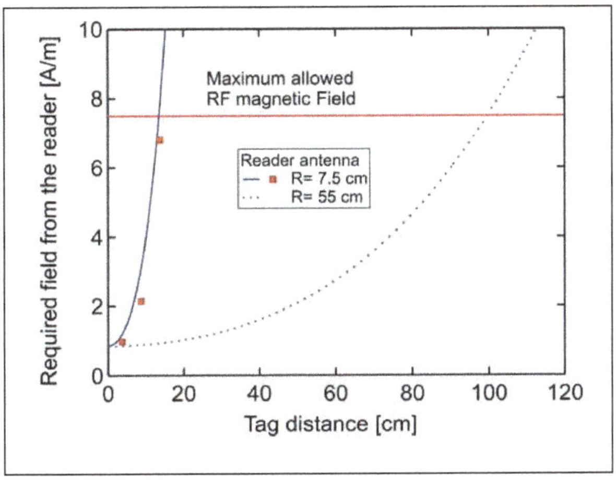

Calculation and experimentally obtained data of the required RF magnetic field at the reader side as a function of the tag distance in order to generate the required RF magnetic field to operate the tag.

Two of the reader standards at 13.56 MHz base carrier frequency are the proximity (ISO 14443) and vicinity readers (ISO 15693). The main difference between them is the coil radius, being 7.5 cm for the proximity reader and 55 cm for the vicinity reader. This results in a maximum readout distance of 10 cm for the proximity and 1 m for the vicinity reader. As mentioned earlier, the standard (ISO 14443) states also that the tag should be operational at an RF magnetic field of 1.5 A/m, which is significantly lower than the maximum allowed RF magnetic field of 7.5 A/m. One can calculate the required magnetic field at the antenna centre in order to obtain the required field to operate the tag. In our case, the required field for an 8-bit organic RFID tag was 0.97 A/m. This is depicted in figure.

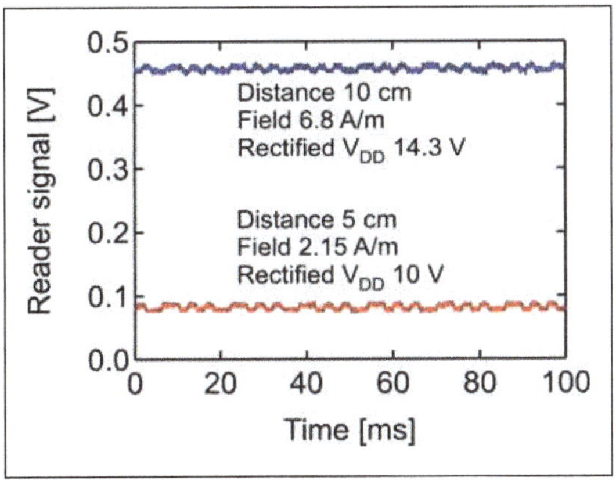

Signal of the 8-bit RFID tag measured on the reader (unamplified reader signal) in DC load modulation mode at distances of 5 and 10 cm.

The dots in this graph show the experimental data at distances of 3.75, 8.75 and 13.75 cm with respect to the field generating antenna. This graph shows that it is possible to energize the 8-bit organic RFID tag at maximum readout distances for proximity readers below the maximum allowed RF magnetic field. The signal detected by the reader during the same experiment is depicted at distances of 5 and 10 cm with respect to the sense coil.

We presented the technology, designs and implementation of an inductivelycoupled passive 64-bit organic RFID tag which is fully functional at a 13.56 MHz magnetic field strength of 1.26 A/m. This RF magnetic field strength is below the minimum required RF magnetic field stated in the ISO standards. The 64-bit transponder chip employs 414 OTFTs. It is internally driven by a supply voltage of 14 V generated by an efficient organic, double half-wave rectifier. The obtained data rate of the organic RFID tag is 787 b/s. Also an 8-bit transponder chip was measured in DC load modulation configuration and could be readout at a distance of 10 cm, which is the expected readout distance for proximity readers.

References

- What-are-rfid-tags, asset-tags: camcode.com, Retrieved 28 February, 2020
- Issues-in-integrated-circuit-design-for-UHF-RFID-4316170: researchgate.net, Retrieved 13 June, 2020
- "Canadian Cattle Identification Agency (CCIA) - Canada's national cattle, animal identification program". www.canadaid.com. Retrieved 25 November 2017
- Active-RFID-active-radio-frequency-identification: internetofthingsagenda.techtarget.com, Retrieved 08 April, 2020
- "Lufthansa Group rolls out electronic bag tags with BAGTAG partnership". Future Travel Experience. January 9, 2018. Retrieved May 1, 2018
- Organic-rfid-tags, radio-frequency-identification-fundamentals-and-applications-design-methods-and-solutions: intechopen.com, Retrieved 28 May, 2020

3

Radio Frequency Identification Middleware

The RFID middleware is a set of components which aims to manage RFID readers, deals with RFID-events and data, and is connected to end-user applications. The middleware is located between readers and business applications. All the aspects related to RFID middleware have been carefully analyzed in this chapter.

RFID Middleware is a radio-frequency identification (RFID) software, that sits between the readers and the enterprise/business applications. The middleware has several functions and plays a major role in RFID system operation and management. The middleware not only manages RFID readers and printers and communicates between these devices and your business applications but also manages, filters, aggregates and makes sense of the data coming from the RFID tags.

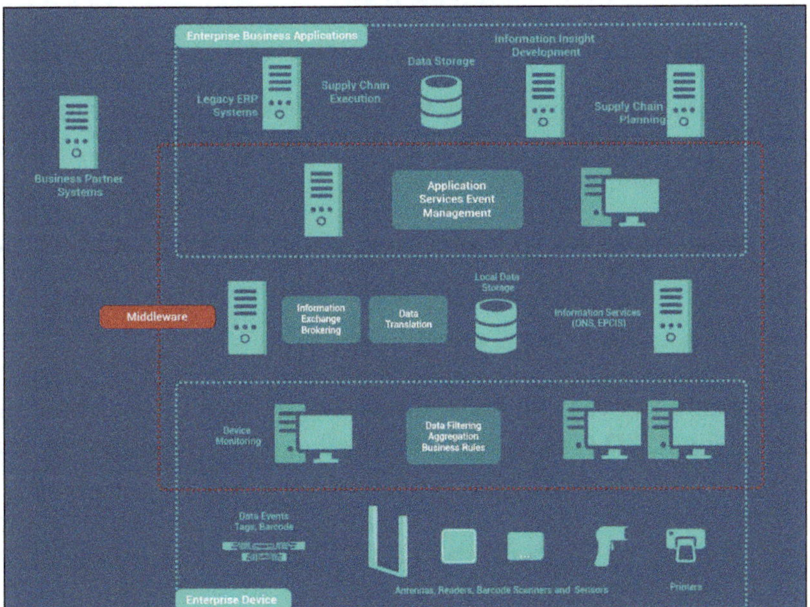

RFID middleware includes extensive data collection and management tools including data collection, filtering and aggregation, and tools for RFID device integration and management.

Due to the sheer amount of data coming from the RFID tags (imagine hundreds or thousands of tags read at virtually the same time thousands of times, radio-frequency identification (RFID) systems require a new kind of middleware, as well as conventional integration middleware. Conventional middleware is used primarily to link disparate applications, both internally and externally, to the enterprise for:

- Routing data using different transport protocols.
- Translating data into different formats.
- Web services.

Requirements for RFID Middleware Function

There are several requirements for RFID middleware such as:

- The ability to efficiently interface with various databases.
- The ability to integrate any AIDC device regardless of manufacturer including RFID readers, RFID printers/encoders, sensors and other devices.
- It has to manage devices from central and distributed locations, and over the cloud and has to be also manageable regardless of system location.
- Has to most importantly interface with various enterprise applications such as Supply Chain Management (SCM), Warehouse Management System (WMS), Enterprise Resource Planning (ERP), etc.
- Having modular architecture, that is scalable and easy to upgrade.
- Providing platform transparency to mask heterogeneity to enable an exchange between disparate platforms, such as different hardware (PC, Mac, Solaris, etc.) and operating systems (Linux, Unix, Windows, Apple). A.
- Support radio-frequency identification (RFID) standards and provide for future standards.

EPCglobal Specifications for RFID Middleware

The EPCglobal community that includes many technology vendors and end users that design and deploy the RFID technology and develop RFID Middleware, has developed specifications that standardize the interfaces between RFID tags, RFID readers and enterprise systems. There are four main standards: Low Level Reader Protocol (LLRP), Discovery Configuration and Initialization (DCI), Reader Management (RM) and Application Level Event (ALE) Standard.

LLRP

The LLRP Standard specifies an interface between RFID Readers and Clients. This protocol is called a low-level protocol because it controls the RFID air protocol operation timing as well as the access to air protocol command parameters.

DCI

The DCI Standard specifies an interface between RFID Readers and Access Controllers and the network on which they operate. The standard is aimed to specify the the operations of a Reader and Client that allow them to utilize the network to which they are connected to communicate with other devices, exchange configuration information, and initialize the operation of each Reader.

RM

The current RM Standard describes the wire protocol used by management software to monitor the operating status and health of EPCglobal compliant RFID Readers. The Reader Management Protocol specifies the interaction between a device capable of interfacing with tags, and management software.

ALE

The ALE Standard specifies a means for clients to specify, in a high-level, declarative way, what electronic product code (EPC) data they are interested in, without dictating an implementation. It provides a standardized format for reporting accumulated, filtered EPC data that is largely independent of where the EPC data originated or how it was processed.

RFID Middleware has an irreplaceable place in the RFID system. Even though sometimes its functions are taken over by the system applications and there is no clear delineation where middleware ends and the user application starts all the tasks are still part of the system and data flow is still handled within.

COMMON FUNCTIONS OF RFID MIDDLEWARE

RFID middleware goes beyond simply connecting devices; rather, it allows users to collect, manipulate, and disseminate data with ease.

- Manage Devices: Middleware is able to control the networking and settings (and other similar devices) and provides a simple way to update any changes that need to be made, even if there are varying types and brands of RFID readers on the network. For example, middleware may include a feature like the ability to monitor the performance level of the readers, which allows users to better predict potential problems and ensures all readers are running at peak performance.

- Collect and Integrate Data: Middleware acts as a bridge, connecting different systems together and enabling the efficient transport of data. When tags are read by an RFID system, the data collected can be sent directly to the middleware in order to process it and then send it where it needs to go (whether that is a warehouse management system, internal database, or some other destination).

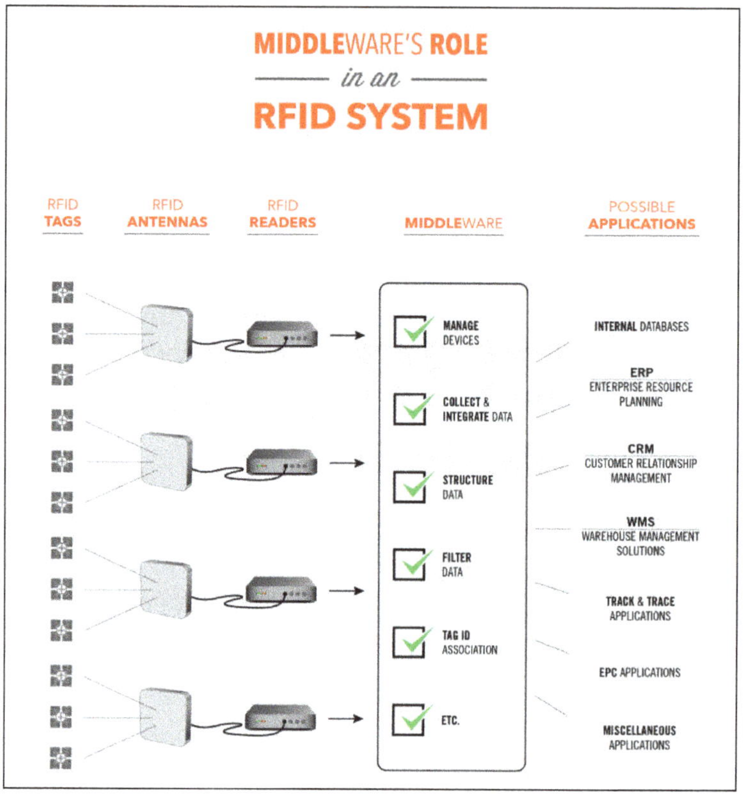

- Structure Data: Middleware provides users the ability to understand what data they are collecting from each device (RFID reader, barcode scanners, etc.) and map out the structure of where that data goes and when. EPCglobal released a software specification in order to streamline and help users create a more detailed and useful structure with middleware for their data. The specification, called Application Level Events or (ALE), is designed to help middleware filter and sort through raw data in order to provide users with relevant information based on "events" that occur. (Events are predetermined by the user as items that are significant for the business.)

  For example, a shipping company wants to know the exact time pallets arrive at the dock door. The first time a tagged pallet is read at the dock door would be an event, and the relevant data is sent to an application. The RFID tags on the pallet are likely read hundreds to thousands of times afterwards, but all of those tag reads can be filtered out by the middleware and not sent or processed by any system.

The ALE specification gives users the freedom to input information such as from which locations to read, how to filter the data, and how to the group the results. Users can then subscribe applications to the structured events allowing the applications to receive reports or messages whenever an event occurs. Applications can subscribe to any number of events and any type of events without affecting one another.

- Filter Data: RFID middleware can be used to filter out the RFID tag reads deemed most important and remove redundant or repetitive reads. In some applications, RFID tagged items are read over and over again by the same RFID reader depending on how long the equipment is stagnant in the read field. RFID Middleware can assist by filtering out hundreds of identical reads so that the user only receives the data important to the application, such as a timestamp of when the equipment was placed inside of and removed from the read field. Filtering out excess data can be extremely helpful for users because massive amounts of tag data can slow down networks and fill up databases quickly.
- Tag ID Association: In a system with hundreds to thousands of RFID tags, RFID middleware can keep track of and allocate tag ID numbers. In a situation where a company is writing hundreds of RFID tags to apply to pallets or cartons of many different types of items, middleware can keep track of which tag ID numbers have been used and assign tags IDs relevant to the item type, date, etc. Middleware can then record the allocation of that number, that the tag was applied properly, and, later, the first read of the new tag.
- Miscellaneous: Because the potential uses for middleware are nearly endless, the same can be said of the potential functionality. Additional integrations, languages, or formats are just a few examples.

RFID MIDDLEWARE DESIGN AND ARCHITECTURE

Radio Frequency Identification (RFID) is a form of Automatic Identification and Data Capture (AIDC) technique. RFID is recently being used in a wide range of areas such as Supply Chain Management (SCM), health care, traffic monitoring, retail, and access control. The ability to store large amounts of data and identify items which are not in the line of sight has given RFID technology an edge over other automatic identification approaches such as the barcode based systems and optical character recognition systems (OCR). As an example, RFID technology integration in SCM systems has resulted in the reduced losses and improved visibility in various stages of supply chaining, reduced numbers of data entry errors, efficient inventory management, and lower human labor costs in distribution centers.

A binary code comprising a field of bars and gaps arranged in parallel configuration is used by the barcode based identification systems. The analysis of the reflected beam on the bar gaps, allows the numerical and alphanumerical interpretation of the barcode sequence made up of narrow and wide bars. The interpreted value obtained specifies a unique code that is used for object identification. The disadvantage of the barcode system is that the barcode needs to be aligned in order to be read by the laser scanner. The OCR based systems consist of optical machine readers used to recognize alphanumeric codes which are placed on the objects to be uniquely identified. The drawbacks of this system consist of the cost of operation, and the complexity of the OCR readers.

The RFID systems basically consist of three elements: a tag/transponder, a reader and a middleware deployed at a host computer. The RFID tag is a data carrier part of the RFID system which is placed on the objects to be uniquely identified. The RFID reader is a device that transmits and receives data through radio waves using the connected antennas. Its functions include powering the tag, and reading/writing data to the tag. As shown figure, the signals sent by the reader‘s antennas form an interrogation zone made up of an electromagnetic field. When a tag enters this zone, it gets activated to exchange data with the reader. Later, the identification data read by the RFID reader is processed by the software system, known as the RFID middleware. The RFID middleware manages readers, as well as filters and formats the RFID raw tag data so that they can be accessed by the various interested enterprise applications. Hence, the middleware is a key component for managing the flow of information between tag readers and enterprise applications.

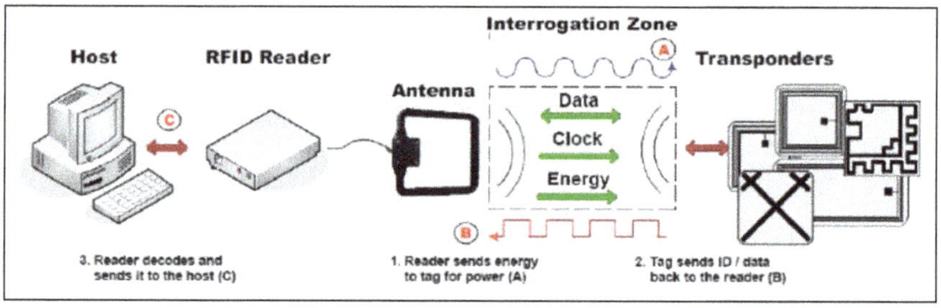

RFID system components.

Major advantages of using RFID as an auto-ID system are the following:

- RFID readers do not require a line of sight to access data from the RFID tags.
- RFID systems can read data over varied range from few centimeters to few hundred meters.
- RFID readers can interrogate, and make RFID tags readings much faster.
- RFID systems can read and write different sizes of data from/to the tag, based on the type of tag.
- RFID systems can read tags in harsh environments, without any human interference.

RFID technology is becoming ubiquitous as RFID systems have recently undergone significant improvements. A variety of makes and models of RFID tags and readers, combined with decreasing RFID hardware prices, are making RFID deployment more attractive. In the traditional applications of RFID such as access control, networking was not a concern and there was barely a need for a RFID middleware solution. However, in the novel application areas such as SCM, a number of RFID readers could be used to capture RFID data which need to be disseminated to a variety of enterprise applications. Hence, there is no longer a one-to-one relationship between reader and application.

The researchers in this area have reported a vast amount of research about the benefits, possible misuses, ethical issues, and technical issues involved in the RFID technology. However, less significant attention has been paid to the issues involved in the RFID middleware that manages large deployments of readers producing high volumes of captured data, and encapsulates applications from the low level data by transforming them into more meaningful events. Considering this void in the RFID middleware research, herewith, we discuss the design issues of RFID middleware, present our solution called FlexRFID which addresses the above aspects, and compare it to other middleware solutions. We analyze FlexRFID to the extent to which it addresses applications' needs, and allows an easy management of devices.

RFID System Components

RFID systems are produced by many manufacturers and exist in countless variants. However, a RFID system consists mainly of three components; the transponder/tag, reader, and RFID middleware.

RFID Transponder/Tag

A RFID transponder, or tag, consists of a chip and an antenna. A chip can store a unique serial number or other information based on the tag‘s type of memory. The tag‘s type of memory can be read-only, read-write, or write-once and read-many. Read-only tags are much cheaper to produce and are used in most current applications. Read-write tags are useful when information needs to be updated. The antenna is used to transmit information from the chip to the reader, and the larger the antenna the longer the read range. The RFID tag can be either attached or embedded in an object to be identified, and can be scanned by mobile or stationary readers using radio waves. RFID tags exist in three different versions: passive tags, active tags, and semi-passive/semi-active tags.

Passive Tags

Present the simplest version of RFID tags which do not contain their own power source, such as a battery, and cannot initiate communication with the reader. The passive tag derives its power from the energy waves transmitted by the reader and responds to the reader‘s radio frequency emissions, therefore the passive tag relies entirely on the reader as its power source. A passive tag should store, at a minimum, a unique identifier for the item tagged, and can be read from a range of about 10 to 20 feet under perfect conditions. Passive tags have lower production costs, meaning that they can be applied to less expensive disposable goods.

The cost of passive tags varies based on the radio frequency used, amount of memory, and design of the antenna, and other tag requirements. Passive tags can operate at low, high, ultrahigh, or microwave frequency. The development of passive RFID tags has made wide scale use of them in many organizations. Examples of passive tag applications include mass transit passes, building access badges, and consumer products in the supply chain.

Active Tags

Unlike passive tags, active tags contain a power source and a transmitter, in addition to the antenna and chip, and send a continuous signal. These tags typically have read/write capabilities; tag data can be rewritten and modified. Active tags can initiate communication and communicate over longer distances up to 750 feet, depending on the battery power. Because these tags contain more hardware than passive RFID tags, they are more expensive and are reserved for costly items that are read over greater distances. RFID manufacturers typically do not quote prices for active tags without first determining their storage type and quantity, and range.

Semi-passive Tags

This type of tags is called also semi-active tags. Semi-passive tags do not initiate communication with the reader but contain batteries that allow the tag to perform other functions, such as monitoring environmental conditions and powering the tag's internal electronics. In order to conserve battery life, some semi-passive tags do not actively transmit a signal to the reader. Instead, they remain dormant until they receive a signal from the reader. Semi-passive tags can be connected to sensors to store information for container security devices. Semi-passive tags have the middle transmission range and cost.

As a summary, passive tags are consequently much lighter than active tags, less expensive, and offer a virtually unlimited operational lifetime. The trade off is that they have shorter read ranges than active tags and require a higher-powered reader (Association for Automatic Identification and Mobility, n.d.). Table shows a comparison among passive, semi-passive, and active tags.

Table: Characteristics of passive, semi passive and active RFID tags.

	Passive Tags	Semi-Passive Tags	Active Tags
On board power supply	No (From Reader)	Yes (Internal Battery)	Yes (Internal Battery)
Transmission range	Short (up to 6.096 meters)	Medium (up to 30.48 meters)	Long (up to 228.6 m)

Communication pattern	Passive	Passive	Proactive
Cost	Cheap	Medium	Expensive
Type of memory	Mostly Read-Only	Read-Write	Read-Write
Life of tag	Up to 20 years	2 to 7 years	5 to 10 years

RFID Tags by Type of Memory

RFID Tags have various types of memory:

- Read-Only tags: Have minimal storage capacity (typically less than 64 bits) and contain permanently programmed data that cannot be altered. These tags primarily contain item identification information and have been used in libraries and video rental stores.
- Read-Write tags: In addition to storing data, they can allow the data to be updated when necessary. Consequently, they have larger memory capacity and are more expensive than read-only tags. These tags are typically used where data may need to be altered throughout a product‘s life cycle, such as in manufacturing or in supply chain management. Read-Write tags have three main procedures for managing and storing data.
- EEPROM (Electrically Erasable Programmable Read-Only Memory): is a type of nonvolatile memory used to store small amounts of data that must be saved when power is removed. It is the most dominant procedure in many RFID systems, but has the disadvantages of high power consumption during the writing operation and a limited number of write cycles.
- FRAM (Ferromagnetic Random Access Memory): Its read power consumption is lower than the EEPROM by a factor of 100 and the writing time is 1000 times lower. Because of manufacturing problems, its widespread introduction onto the market was affected.
- SRAM (Static Random Access Memory): SRAM are used for data storage in microwave system which facilitate very fast write cycles. The disadvantage of this procedure is that the data requires an uninterruptible power supply from an auxiliary battery (active transponder).
- Write-Once, Read-Many tags: Allow information to be stored once, but does not allow subsequent updates to the data. This tag provides the security features of a Read-Only tag while adding the additional functionality of Read-Write tags.

RFID Tags Operation Frequencies

RFID tags operate in several frequency bands. Most of the used frequencies are those that are in the Industrial, Scientific or Medical (ISM) frequency ranges.

RFID frequencies are divided into the following three basic ranges:

- Low Frequency (LF): This range operates between 30 and 500 KHz. However 125-134 KHz is the most ordinary range used in animal tracking, car immobilizers, security access, asset tracking etc. LF tags are commonly used where there are liquids, electrical noise, or metals present and when a fast read rate is not required. Most of low frequency systems operate without the need of integrated battery in their tags, have short reading ranges, and are lower system costs.
- High Frequency (HF): This range operates between 10-15 MHz, but 13.56 MHz HF tags are the most commonly used, due mainly to the relatively wide adoption of smart cards based on RFID technology. The cost of the high frequency systems is inexpensive, but higher than the low frequency systems, they have longer read ranges and higher reading speeds than the LF systems. The HF systems are used in access control and smart cards.
- Ultra High Frequency (UHF) and Microwave Frequency: Ultra High Frequency Systems operate between 400 and 1000 MHz and microwave frequencies between 2.4 and 2.5 GHz. These systems are the most expensive compared to the others. UHF tags are considered as being the most practical for item-level tracking as they offer a good balance between range (typically less than a few meters), a high reading speed, and the ability to read multiple tags. Unlike the other systems, line of sight is required for the communication between RFID reader and tags. UHF systems have a very long read range, and are used for such applications as railroad car tracking and automated toll collection. Microwave frequency band is also used by many other systems e.g. Bluetooth and Wi-Fi systems.

RFID Reader

A RFID Reader is a scanning device that reliably reads the tags and communicates the results to the middleware. A reader uses its own antennae to communicate with the tag by broadcasting radio waves to which all tags within range will respond. Readers can process multiple items at once, allowing for increased read processing times. They can be either mobile or stationary, and they are differentiated by their storage capacity, processing capability, and the frequency they can read.

RFID reader consists of the following functional blocks:

HF Interface

The master part of the reader which has these functions:

- Supplying RFID transponders with power by generating high frequency power.
- Modulation of the signal to the transponder.

- Reception and demodulation of signals from the transponders.

Control Unit

The slave part of the reader that performs the following functionalities.

- Communication and execution of the application software‘s commands.
- Signal coding and decoding.
- Communication control with a transponder.

Some RFID readers have additional functionalities like anti-collision algorithm, encryption and decryption of transferred data, and transponder-reader authentication. Different designs of readers exist, because different applications have different requirements from each other. RFID readers are classified into three types:

- OEM readers: Original Equipment Manufacturers readers are mostly used for data capture systems, access control systems, and robots.
- Industrial use readers: used in assembly and manufacturing plant.
- Portable readers: These readers are more mobile than the other readers, and supported with a LCD display and keypad. This kind of readers is used in animal identification, device control and asset management applications.

RFID Middleware

The middleware refers broadly to software or devices that connect RFID readers and the data they collect, to enterprise information systems. RFID middleware helps making sense of RFID tag reads, applies filtering, formatting and logic to tag data captured by a reader, and provides this processed data to back-end applications. RFID middleware serves in managing the flow of data between tag readers and enterprise applications, and is responsible for the quality, and therefore usability of the information. It provides readers connectivity, context-based filtering and routing, and enterprise / B2B integration. RFID middleware design and components will be discussed further in the next sections.

When designing a RFID middleware solution, the following issues need to be considered:

- Multiple hardware support: The middleware must provide a common interface to access different kinds of hardware offering different features.
- Synchronization and scheduling: There should be intelligent scheduling and synchronization among all the processes of the middleware. This minimizes the latency and improves the efficiency of the middleware.

- Real-time handling of incoming data from the RFID readers: The middleware should handle the huge amount of data captured by the connected readers in real time without read misses.
- Interfacing with multiple applications: The middleware should be capable of interacting with multiple applications simultaneously, by catering to all the requirements of the applications with minimal latency.
- Device neutral interface to the applications: The application developer should only use the generic set of interfaces provided by the middleware independently of the type of hardware connected to the system.
- Scalability: The middleware design must allow easy integration of new hardware and data processing features.

RFID Middleware Components

A RFID middleware is the interface that sits between the RFID hardware and RFID applications. It provides the following advantages:

- It hides the RFID hardware details from the applications.
- It handles and processes the raw RFID data before passing it as aggregated events to the applications.
- It provides an application level interface for managing RFID readers and querying the RFID data.

A layer of the RFID middleware incorporates all the device drivers of different hardware and exposes to the application standard interfaces to access this hardware. If the application was provided with all the device drivers of all connected readers, it will be a hard job to manage and interface each of the devices. The application developer will then need to understand all the hardware specific internals and operations. Also, the application, if provided with the huge amount of raw tag data reported by the readers, will find it very difficult to process the data in real time. A RFID middleware provides a standardized way of dealing with this flood of information, which processes the raw data and provides the application with clean and filtered data.

A RFID middleware is generally composed of four major layers:

- Reader Interface.
- Data Processor and Storage.
- Application Interface.
- Middleware Management.

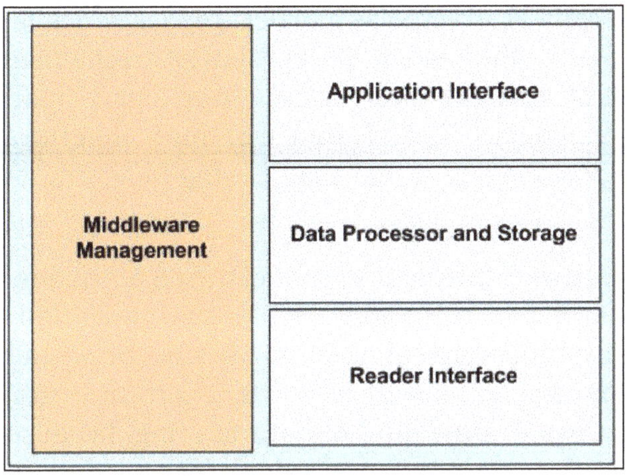

RFID middleware components.

Reader Interface

The reader interface is the lowest layer of the RFID middleware which handles the interaction with the RFID hardware. It maintains the device drivers of all the devices supported by the system, and manages all the hardware related parameters like reader protocol, air interface, and host-side communication.

Data Processor and Storage

The data processor and storage layer is responsible for processing and storing the raw data coming from the readers. Examples of processing logic carried by this layer are data filtering, aggregation, and transformation. This layer also processes the data level events associated with a specific application.

Application Interface

The application interface provides the application with an API to access, communicate, and configure the RFID middleware. It integrates the enterprise applications with the RFID middleware by translating the applications' requests to low level middleware commands.

Middleware Management

The middleware management layer helps managing the configuration of the RFID middleware, and provides the following capabilities:

- Add, configure, and modify connected RFID readers.
- Modify application level parameters such as filters, and duplicate removal timing window.
- Add and remove services supported by the RFID middleware.

RFID readers are typically abstracted as a logical reader which is either a collection of several readers or a part of the reader. This grouping mechanism is used where there is a need to have a set of readers capturing data from a particular area such as a warehouse with many loading docks. The advantage of this is that the application can query a small number of logical readers rather than having to aggregate events from each of the individual readers.

There are two standardized interaction models used to define the communication between the middleware and the applications. An application can operate at synchronous mode when requesting services on demand or asynchronous mode when it registers for information to be sent to it when certain conditions are met. RFID middleware usually provide some kind of data filtering, because sometimes it might be required to report only certain type and value of the tag data to the application. The application needs to provide a set of defined patterns to the middleware. The middleware then allows only data that matches the pattern to be reported to the application. E.g. if an application needs to see only tag data that starts with a specific pattern such as "XYZ20", the filter can be set to this value by the application and communicated to the middleware.

EXAMPLES OF RFID MIDDLEWARE SOLUTIONS

The Savant Middleware

There have been some proposals and research work involving middleware design and RFID data processing. The Auto-ID Center has developed a middleware component called Savant that collects, accumulates, and processes Electronic Product Code (EPC) data obtained from several RF readers. It adjusts multiple readings of a tag, and performs tasks such as archiving data, and inventory control.

The Savant has a set of Processing Modules or Services which may be combined to meet the user's application's needs. This modular structure allows innovation to be promoted by independent groups of people, which helps avoiding the creation of a single monolithic specification that attempts to satisfy all needs for everybody. Figure shows the three key elements of the Savant middleware architecture: Event Management System (EMS), Real-Time in-Memory Data Structure/Real-Time in-Memory Event Database (RIED) and Task Management System (TMS).

The EMS provides a JAVA API for different types of RF readers and it serves to collect tag read events. The EMS allows adapters to be written for various types of readers, collecting EPC data from readers in a standard format, allowing filters to be written to smooth or clean EPC data, allowing various loggers to be written, and buffering events to enable loggers, filters and adapters to operate without blocking each other.

The EMS is composed of the following elements:

- Reader Interface: Allows readers and adapters to communicate events detected by the Auto-ID readers.
- Reader Adapters: Communicate with readers to capture EPC events.
- Event Loggers (Event Consumers): Allow for varied processing of events; store the information in the database, store events in a memory data structure, and broadcast the events to remote servers.
- Event Queues (Event Forwarders): Handle multiple reader event loggers with synchronous implementations.

The RIED is an in-memory database that can be used to store event information by Edge Savants. It provides the same interface as a database, but offers much better performance. The RIED should be a high-performance in-memory and a multi-versioned database.

The TMS manages tasks, just as the operating system manages processes, and provides an interface for task management. Task examples include data gathering, remote task scheduling, personnel alerts, and remote upload. The TMS should be a platform-independent system requiring little memory processing power, should automatically upgrade the tasks it executes, and should present a well-defined, interoperable external interface to schedule, monitor, and remove tasks. Tasks should also be written in a platform-independent language using a simple well-defined SDK.

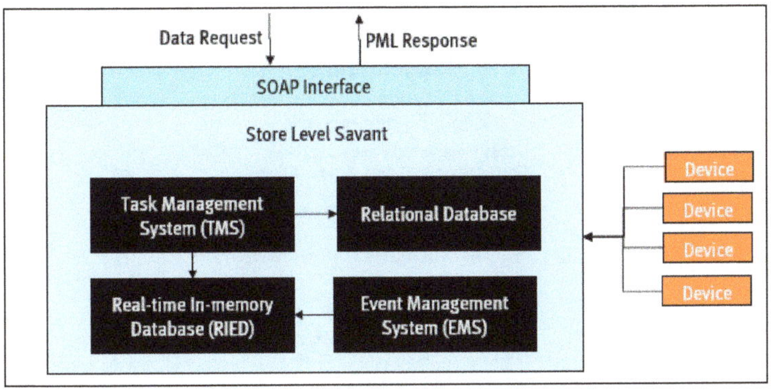

Savant middleware key components.

WinRFID Middleware

WinRFID developed at the University of California Los Angeles (UCLA), is another middleware architecture that uses web services and enables rapid RFID applications development. It is a multi-layered middleware that consists of five main layers shown in figure.

The physical layer deals with the hardware consisting of readers, tags and other sensors. This layer abstracts the hardware elements; readers, tags and host I/O interfaces.

This abstraction allows extending the middleware capabilities in the advent of introduction of new RFID technology.

The protocol layer: The ability to support multiple tag protocols and add new ones is becoming imperative in middleware designs. The protocol layer of the WinRFID middleware allows abstracting the reader-tag protocols. It wraps the command syntax and semantics of a variety of published protocols such as ISO 15693, ISO 14443, ISO 18000–6 A/B, ICode, EPC Class 0 and EPC Class 1. It also deals with protocol specifics such as byte-based, block or even page reading and writing, structure and length of the command frames, partitioning of the tag memory space, checksums, etc.

The data processing layer deals with processing data streams generated by the network of readers. It includes processing rules that deal with problems due to tag density, read/write distance, orientation of tags and material of item that introduce inconsistencies in reading or writing such as multiple reads of the same tag, some tags not being read, erroneous reads, etc. All of these discrepancies are processed as exceptions and a variety of altering systems are available for resolution such as emails, messages, and user defined triggers.

The XML framework layer formats the cleaned tag data in a variety of ways to a higher level XML based representation. The purpose of this layer is to provide data in a suitable format to the application layer for decision making.

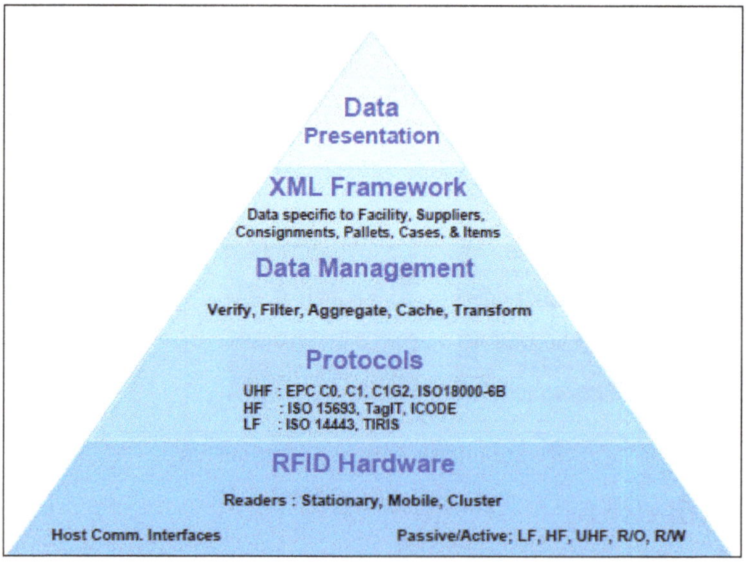

WinRFID middleware multi-layered architecture.

The data presentation layer presents the data as per the requirements of end-users or different applications requirements. It facilitates data visualization for decision making. This layer supports two components the portal and the database connector. The portal provides the users with an interface to subscribe to the information of interest. For the database connector, currently the middleware can populate SQL Server and

Oracle RDBMS. The databases get populated in an asynchronous fashion in a trickle mode – a process with least priority so as to avoid the edge hosts getting locked up.

WinRFID exploits the .Net framework's runtime plug-in feature to support the addition of new readers, protocols, and data transformation rules with minimum disruption of the existing infrastructure.

The WebSphere RFID Middleware

The WebSphere RFID middleware solution, designed by IBM, consists of three main components as shown in figure. RFID devices, WebSphere Premises Server, and Websphere Business Integration Server.

The IBM WebSphere is a sensor enabled product that allows sensor data aggregation and analysis, deriving insights from sensor data and integrating those insights with the SOA business processes. The software provides the use of intelligent business rules that manage complex event identification and processing.

This solution expands device services allowing a single platform to support multiple sensor types, and supports workflow tooling for sensor data integration with business processes. Therefore, it delivers new and enhanced capabilities to create a robust, flexible, and scalable platform for capturing new business value from sensor data.

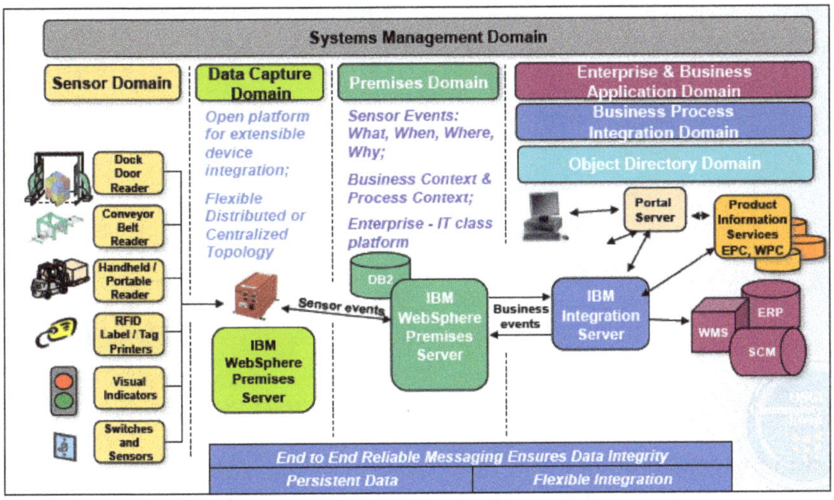

The IBM sensor and actuator solutions framework.

The Sun JAVA RFID System

Sun Java System RFID software is a Java based commercial middleware platform provided by Sun. It is a critical RFID infrastructure component that allows a safe, secure, and efficient data and device integration from the edge of the enterprise into enterprise application systems. It has a dynamic, service provisioning architecture that enables scaling from small pilots to large deployments with high data volume.

The Java System RFID Software supports a variety of new and existing standards, such as EPC, ISO, Gen 2, passive and active tags and devices, read/write tags, and commercial and government standards. It is a part of the Java Enterprise System (JES) and has four components as shown in figure: the RFID Event Manager, the RFID Management Console, the RFID Information Server, and a Software Development Kit (SDK). The RFID Event Manager is a Jini-based event management system that facilitates the capture, filtering, and eventual storage of events generated by RFID readers. The RFID Management Console provides a browser based management interface, which allows configuration of various attributes and parameters of the middleware. The RFID Information Server is responsible for storing and querying the EPC related data, it also manages inter Enterprise handling of the data. The SDK provides a development platform to build custom applications.

The Sun middleware exposes to the application, the hardware as logical readers. These logical readers may be a collection of one or more physical readers that the application can select and apply the various processing parameters to the group.

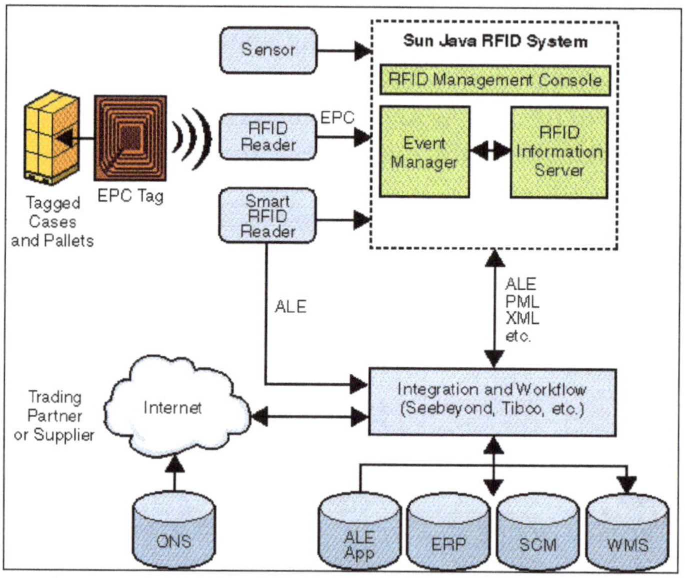

The Sun Java RFID system function in the EPC network.

All of these middleware designs aim at providing a scalable solution for gathering, filtering, and providing clean RFID data to the end-user. However, there are still many open issues. The reliability of RFID data needs to be improved since inaccurate data could misguide the application users. The accumulation of RFID data generated in high volumes, may lead to slower queries and updates, therefore efficient RFID data management solutions such as data transformation, aggregation, and dissemination should be investigated. Raw RFID data is not of significant value until it is aggregated with other data to obtain appropriate inferences, and transformed into a suitable form for application level interaction. Also, the applications with high security requirements are increasingly using RFID; therefore support for data security and confidentiality is needed.

However, such support should maintain a desirable system performance. RFID also raises the privacy concerns because of its potential to leak proprietary information and ability to track private information such as the spending history of a consumer. Technical solutions must be implemented to ensure that private data is not compromised with.

While the Savant middleware architecture provides features for cleaning the data and interfacing with different kinds of RF readers, it has limited built-in functionality for addressing business rules management, dealing with all types of sensor devices and providing data dissemination, filtering, and aggregation. Also, none of WinRFID and IBM WebSphere considers the business rules policies implementation, especially the ones concerned with security and privacy.

The FlexRFID design aims to provide the applications with a device neutral interface to communicate simultaneously with many different hardware devices, creating an intelligent RFID network. It also provides an interface to access the hardware for the management and monitoring purposes. The FlexRFID provides all data processing capabilities along with the security and privacy features included in the data processing layer and enforced by a policy based management module for the business events, referred to as the Business Rules layer. The modular and layered design of FlexRFID allows integration of new features with little effort. The design also permits seamless integration of different types of enterprise applications.

FLEXRFID

FlexRFID is part of a three-tier architecture consisting of: the backend applications layer, FlexRFID middleware layer, and hardware layer consisting of diverse types of sensors and devices.

The Diverse Types of Sensors and Devices layer comprises RFID readers, sensors and other industrial automation devices. Such approach allows incredible flexibility in the selection of devices, lets companies build their enterprise solutions without handling low-level programming, and allows creating an intelligent sensor network, where RFID readers are choreographed with other devices. There are diverse makes and models of devices, which require a middleware layer that monitors, manages, coordinates, and obtains data from the different devices. In FlexRFID, these functions are taken care of before processing the raw data and applying business logic to them. Our approach is to use a Device Abstraction Layer (DAL) that abstracts the interaction with the physical network of devices. The FlexRFID middleware incorporates three other layers which are: Business Event and Data Processing Layer (BEDPL), Business Rules Layer (BRL), and Application Abstraction Layer (AAL).

Device Abstraction Layer (DAL)

The Device Abstraction Layer of the FlexRFID middleware is responsible for interaction with various devices and data sources independently of their characteristics. The Data Source Abstraction Module (DSAM) of the DAL provides a standard view of data regardless of the data source protocol (e.g. EPC Gen2, ISO 15693, and ISO14443A), air interface (e.g. UHF, HF), power supply, type, and memory size of a device. The Device Abstraction Module (DAM) of the DAL provides a common interface to access hardware devices with different characteristics such as protocols, air interface, and host-side communication interface (e.g. USB, Serial Port, Ethernet port). The DAM exposes simple functions like open, close, read, write, etc. that trigger the complex operations of the devices. Both, the DSAM and the DAM allow the FlexRFID middleware to be extendable to support various data sources and devices. The Device Management and Monitoring Module (DMMM) of the DAL is responsible for dynamic loading and unloading of the driver libraries or device adaptors. This allows the FlexRFID middleware to be light weight as libraries are loaded based upon request. The DMMM configures the devices as specified by the upper layers, and also monitors and reports their status.

Business Event and Data Processing Layer (BEDPL)

The BEDPL acts as a mediator between the DAL and the AAL. The services accepted by the BEDPL are first authorized by the Business Rules Layer (BRL) and then allowed to issue commands to the DAL in order to get the raw data and process them accordingly. Similarly the raw data are carried from the DAL, processed, and passed on to the AAL by this layer. Services provided by the BEDPL are:

Data Dissemination

A diverse set of applications across an organization are interested in the captured information. The captured data are therefore broadcasted by the data dissemination service to all the interested entities. In addition, different applications require different latencies. For example, low latency for the notifications is desired by the applications that need to respond immediately to objects' events. In contrast, some legacy applications need to receive batched updates on a daily schedule.

Data Aggregation

The fine-grained data has implicit meanings and associated relationships with other data, and need to be aggregated into summaries and proper inferences for applications that can not deal with the increased granularity. For example, it is common that an application is only interested in an event when an object enters or leaves a certain area. Other applications may only need a total count of objects belonging to a specific category rather than a serial number of each object detected. The data aggregation service provides such kind of functionality.

Data Transformation

Raw data present little value until they are transformed into a form suitable for application-level interactions. So, from an application perspective, it is desirable to provide a mechanism that turns the low-level captured data into the corresponding business event.

Data Filtering

The volumes of data generated by the different devices require significant data filtering to extract the most important information. Also, different applications are interested in different subsets of data captured. There are filtering policies available in the FlexRFID middleware policy repository of the BRL, therefore the data filtering service filters data depending on the filter characteristics provided by the application. This offers flexibility in handling multiple filtering formats.

Duplicate Removal

Multiple devices may generate duplicate readings of the data, for example tags in the vicinity of a RFID reader are read continuously. This results in a large amount of repeated data, and therefore duplicate removal service prevents the reporting of these duplicate data. The application specifies a time window, so that the same data read within it are only reported once.

Data Replacement

Usually the rate at which the devices insert data in the channel buffer is slower than the read rate of the applications. However, in case the application is not responsive enough or not executing, the channel buffer gets full, and leads to buffer overflow problem. The data replacement service allows the application to specify the action to be taken in case of channel buffer overflow. The application specifies the data replacement policy stored in the BRL policies repository, which will be executed by the data replacement service.

Data Writing

Certain special data sources like RFID tags provision additional memory space for both ID and additional data. The FlexRFID middleware handles both the reading and writing of data to this additional memory.

Privacy

RFID based tracking solutions could trigger RFID tags attached to the personal belongings to reply with their ID and other private information, therefore increasing the potential of unauthorized surveillance mechanism that would pervade large parts of our lives. FlexRFID design supports dedicated privacy enhancing feature through the privacy module. The business rules of this module are stated in the privacy policy of the BRL.

Business Rules Layer (BRL)

The BRL is a policy-based management engine that defines the rules that grant or deny access to resources and services of the FlexRFID middleware, and enforces different types of policies for filtering, aggregation, duplicate removal, privacy, and different other services. This is achieved by determining the policies to apply when an application requests the use of a service in the BEDPL. The Middleware Policy Editor (MPE) allows storing, retrieving, and removing policies from the Middleware Policy Repository Database (MPRD). When an application needs to access a service that is protected by the Business Rules Layer, the request passes through the Middleware Policy Enforcement Point (MPEP) which asks the Middleware Policy Decision Point (MPDP) whether to permit or deny access to the service by applying the privacy rule, and how the service will be processed depending on its type. The MPEP gives the MPDP the authority of decision making; whether or not to grant the application access to the service based on the description of the application attributes, and which policies will be applied to the services used by this application. The MPDP makes its decision based on the applicable policies stored on the system. The returned decision is Permit, Deny, Indeterminate or Not Applicable. Indeterminate is returned when there is an error in processing the request and Not Applicable when no policy that applies to the request could be found. Policies are operating rules used to maintain order, security, consistency, or other ways of successfully achieving a task. Examples of policies that should be available in the Business Rules Layer are: Access policy, data replacement policy, quality of service policy, and privacy policy.

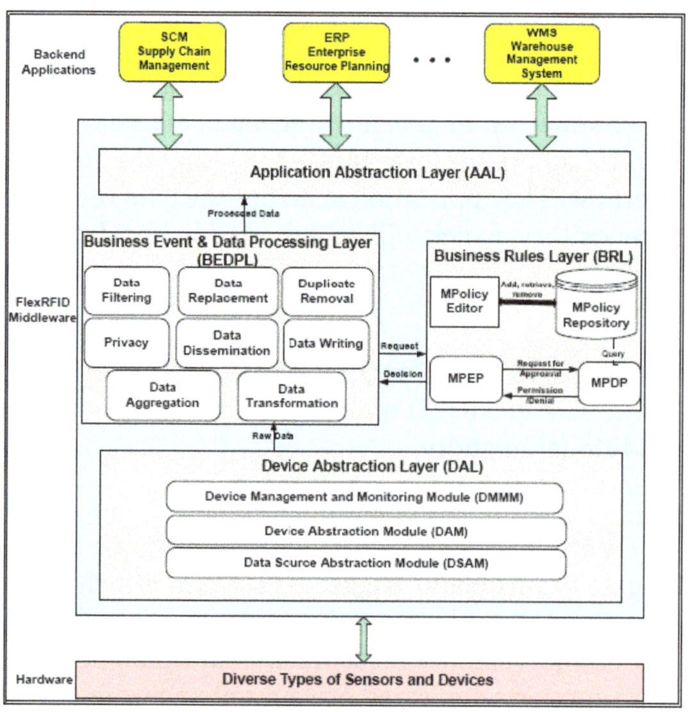

FlexRFID middleware architecture.

Different types of applications using the FlexRFID middleware may define rules to detect events and process them using the services provided by the middleware. Primitive events such as observations from readers may lead to actions such as change of location. Sequence events consist of a sequence of primitive events of the same type, defined by the order and closeness of intervals. Composite events are a combination of primitive events and sequence events, and may lead to actions such as aggregation of data:

- The filtering rule filters data according to predefined policies by the applications. For example, multiple readers may generate duplicate readings. To filter this, the filtering policy will scan data within a sliding window to find if there are duplicate RFID tag readings from multiple readers, and delete the duplicate if it exists. A policy for duplicate removal could specify that if readings from reader Rx and Ry have the same tag ID value within time T, then one of them is dropped.
- The location transformation rule serves to transform RFID readers' observations into location changes. For example, Reader R1 is mounted at a warehouse departure zone and will scan objects before their departure. A policy for this transformation could state that any observation generated from reader R1 will change the object's location to a value different from its current location.
- The data aggregation rule is used to detect a sequence of ordered events and generate an aggregation relationship. For instance when pallets are loaded into a truck to depart, a sequence of readings on the pallets are done, followed by (with a distinctive distance) a separate reading of the truck's EPC. This sequence of events will aggregate as a containment relationship between the pallets and the truck.
- Privacy threats in an RFID application can include covert reading, tracking over time, and individual profiling. The privacy rule specifies whether an application has the right to access RFID tag data, can track them over time, and use them to generate events. Applications can load into the FlexRFID middleware's Business Rules Layer privacy policies specifying how to use and configure the RFID technology to maintain the privacy of data and prevent data from tracking and hotlisting.

Application Abstraction Layer (AAL)

The AAL provides various applications with an interface to the hardware devices, through which the applications request the set of services provided by the FlexRFID middleware with hidden complexity.

4

Issues and Challenges

There are various security, privacy, ethical and technology related problems with RFID. It includes RFID tag collision, RFID reader collision, zombie RFID tags, etc. This chapter has been carefully written to provide an easy understanding of these issues and challenges in radio frequency identification.

SECURITY AND PRIVACY ISSUES

TAG Data

RFID tags are considered "dumb" devices, in that they can only listen and respond, no matter who sends the request signal. This brings up risks of unauthorised access and modification of tag data. In other words, unprotected tags may be vulnerable to eavesdropping, traffic analysis, spoofing or denial of service attacks.

Eavesdropping (or Skimming)

Radio signals transmitted from the tag, and the reader, can be detected several metres away by other radio receivers. It is possible therefore for an unauthorised user to gain access to the data contained in RFID tags if legitimate transmissions are not properly protected. Any person who has their own RFID reader may interrogate tags lacking adequate access controls, and eavesdrop on tag contents.

Researchers in the US has demonstrated a skimming attack on an RFID credit card, through which credit card information, such as the cardholder's name and account information, could be skimmed if not properly encrypted.

Traffic Analysis

Even if tag data is protected, it is possible to use traffic analysis tools to track predictable tag responses over time. Correlating and analysing the data could build a picture of movement, social interactions and financial transactions. Abuse of the traffic analysis would have a direct impact on privacy.

Spoofing

Based on the data collected from eavesdropping or traffic analysis, it is possible to perform tag spoofing. For instance, a software package known as "RFDump",11 that runs on a notebook computer or personal digital assistant, allows a user to perform reading or writing tasks on most standard smart tags if they are not properly protected. The software permits intruders to overwrite existing RFID tag data with spoof data. By spoofing valid tags, the intruder could fool an RFID system, and change the identity of tags to gain an unauthorised or undetected advantage. One example is trying to save money by buying expensive goods that have had their RFID price tags spoofed to display cheaper prices.

By combining the two capabilities of eavesdropping and spoofing, a replay attack is possible where an attacker can "query a tag, receive the information it sends, and retransmit this information at a later time".

Denial of Service Attack

The problems surrounding security and trust are greatly increased when large volumes of internal RFID data are shared among business partners. A denial of service attack on RFID infrastructure could happen if a large batch of tags has been corrupted. For example, an attacker can use the "kill" command, implemented in RFID tags, to make the tags permanently inoperative if they gain password access to the tags. In addition, an attacker could use an illegal high power radio frequency (RF) transmitter in an attempt to jam frequencies used by the RFID system, bringing the whole system to a halt .

RFID Reader Integrity

In some cases, RFID readers are installed in locations without adequate physical protection. Unauthorised intruders may set up hidden readers of a similar nature nearby to gain access to the information being transmitted by the readers, or even compromise the readers themselves, thus affecting their integrity. Unauthorised readers may also compromise privacy by accessing tags without adequate access controls.

As a result, information collected by readers and passed to the RFID application may have already been tampered with, changed or stolen by unauthorised persons. An RFID reader can also be a target for viruses. In 2006, researchers demonstrated that an RFID virus was possible. A proof-of-concept self-replicating RFID virus was written to demonstrate that a virus could use RFID tags to compromise backend RFID middleware systems via an SQL injection attack.

Personal Privacy

As RFID is increasingly being used in the retailing and manufacturing sectors, the widespread item-level RFID tagging of products such as clothing and electronics raises public concerns regarding personal privacy. People are concerned about how their data is being used, whether they are subject to more direct marketing, or whether they can be physically tracked by RFID chips. If personal identities can be linked to a unique RFID tag, individuals could be profiled and tracked without their knowledge or consent.

For instance, washing clothes tagged with RFID does not remove the chips, since they are specially designed to withstand years of wear and tear. It is possible that everything an individual buys and owns is identified, numbered and tracked, even when the individual leaves the store, as far as products are embedded with RFID tags. RFID readers can detect the presence of these RFID tags wherever they are close enough to receive a signal.

Since RFID remains an emerging technology, the development of industry standards for protecting information stored on RFID chips is still being explored and strengthened. Research into the development and adaptation of efficient hardware for cryptographic functions, symmetric encryption, message authentication codes and random number generators will improve RFID security. In addition, advances in RFID circuit design and manufacturing technology can also lower development costs releasing more resources in tags that can be used for other functions, such as allocating power consumption towards security features.

Today, certain public key technologies are also being studied and in some cases deployed by RFID vendors. This helps improve confidentiality, user authentication and privacy of RFID tags and associated applications. RFID vendors are also conducting research into integrity and confidentiality issues around RFID reader infrastructure. Data can now be stored on a token using dynamic re-keying, where specific readers can rewrite a token's credentials/signature, and verify the token's identity. However, the cost and performance issues around using public key technologies in RFID applications have stalled its use for critical security applications.

TECHNOLOGY RELATED ISSUES

Problems with the Standards of RFID

RFID built in different ways by companies have designed, international standards are still under development and interoperability, which is working to achieve well. Here, it should be pointed out that some of the RFID devices do not have to leave the network and the RFID cards used for inventory control within the company.

There are development standards that operate in high frequency and low, but in fact, big companies want to use cards that operate on high frequencies that provide a wide range compared with those working in low frequencies, but in fact Quarry cards that operate in the high frequencies a list price of about $ 1000 or more, large companies need to be thousands of these readers to cover all the required area, and also the need of millions of cards that are inexpensive, but it will be hung on the products be priced a few dollars, which makes the process expensive.

The RFID Systems can be Easily Disrupted

RFID systems that use the electromagnetic spectrum (Wi-Fi networks as well as cell phones), leading to a collision when you are working on the same frequency and to a lot of delays and inconvenience to consumers who want to Pay and get out of the store.

RFID Reader Collision\Interference

Reader collision occurs when the overlap with readers my reference each other, the card is able to respond to the two systems should avoid these situations, so the protocols and found the anti-collision, which enables the card to take the decision to send the information to the reader.

RFID Tag Collision

Occur when there are many of the cards found in a small area, in addition to the reading time is short, so it is easy for vendors to develop systems that ensure a response card and only one by employing the appropriate algorithms.

ETHICAL ISSUES RELATED TO RFID

Tracking of Merchandise

Using RFID for merchandise only raises one primary issue of its use. This issue is the privacy of the individual consumer and their respective information. RFID tags on merchandise are used to track inventory, stolen items, and shopping trends. The ethical issues come into play when items are tracked to the individual person or when an individual's shopping trends are all tracked via RFID. This issue, while minimalistic in intentions, can cause malicious use if the wrong hands can obtain the information. This information, which is normally just used for companies to understand trends in their product purchases, can be obtained and used by others in order to have information to embarrass or defame a specific person who made a purchase. Potentially post sale RFID tags could be used to track where consumers are moving by tracking the objects with RFID tags. Since RFID tags are quite open in terms of technology, anyone with a scanner could perform such tracking. This causes an even bigger possible issue where burglars could potentially remotely search the houses of potential victims by looking for items such as expensive electronics. Part of the danger of RFID tags is that consumers may be carrying these devices without knowing or understanding how to remove them or what these tags do.

People and Animals

RFID Implants and RFID tags in consumer items has sparked controversy in some privacy conscious people. They worry mostly for two things, their physical privacy, and the privacy of their information. Worries are that even when a person is in their home, outside of the public eye, where they maintain a reasonable expectation of privacy, that they can be tracked. Ostensibly, a person with RFID in his body or perhaps his credit card could be tracked wherever he is, and his information could be gleaned by a reader tuned to the proper frequencies, which, due to their widespread use, are known and easily found on the internet. The signal (or lack thereof) can also be a telltale sign of whether or not a person is at home based on the presence of a response from the tag, possibly giving an intruder an opportunity to break in. Luckily for now the distance and penetration of the RFID signal is still relatively short, keeping security concerns down at this time.

RFID can also be used to track animals as well. Pet owners will have their pets implanted with a RFID chip as below with identifying information that would help a pet be returned to its owner, allowing it to be identified even if a pet's collar were to be removed by some outside factor. A major problem associated with pet RFID usage early in its lifetime was the fact that not all implanted chips operated on the same frequency, preventing tragedies like in March 2004 when a dog was euthanized in Virginia because the shelter did not have the proper reader for the dog's tag. However the worry again is that the information that these chips contain, namely about the owner of the animal, which could be used as personally-identifiable, and assist in identity theft or financial fraud.

Implantation

One potential application of RFID is human and animal implantation. An RFID chip is permanently injected into the skin, such as under the wrist of a person. Currently, this is a very rare practice in humans. In animals, it is frequently used for many scenarios. It is common for pet owners to have chips implanted in their pets to aid in recovery of lost animals. Outside of the home, animal implantation is a very useful tool for tracking livestock.

However, when the issue of human implantation is raised, so are a host of ethical issues. Probably the scariest is the possibility that a government would require all citizens to be implanted with a chip, possibly at birth. This would naturally be a gross violation of personal rights. For one, it would make tracking, such as through tags on products, ubiquitous and impossible to avoid. It also violates basic human rights to have control over your own body, and what goes in it. Even without a government mandate, ubiquitous implantation is still a possibility. If more and more every-day interactions, such as purchasing groceries, require this type of implantation, one would have little choice, lest they lose the ability to make purchases altogether.

CHALLENGES IN RFID TECHNOLOGY

Cost

Today, based on the behavior of customers, the history or record of the production including the delivering is not the most important thing. Most consumers care about the price, expiration date, or packing of the goods. In other words, although the RFID can enhance the management of logistic, to embed the RFID technology into current system also cost a lot. For example, one RFID tag for paste once (non-reusable) may cost $0.8. However, a bottle of water may also cost about $1. Most consumers would not to pay almost twice payment to obtain the information which they do not care about. Comparing with barcode or 417-barcode, the RFID tag costs more than barcode or 417-barcode which can be printed by a printer. In other words, to identify an object via RFID tag costs more than using barcode.

In opposition to once-using RFID tag, to reuse the RFID tag may be a solution. A plastic card where RFID embedded can be used for identification or payment. For example, in Taiwan, most mass transit systems can be paid according to the RFID card that pre-registered and sold to the consumers. Each consumer can use the RFID card to pay the consumption in convenient store, fee of the parking lot, and use the card as a ticket for railway or MRT system. Due to the convenience, most consumers have at least one RFID card for payment. However, the security for financial application is very important. Although there are some security method proposed, no algorithm or system can provide the 100% promise that the security method is always safe.

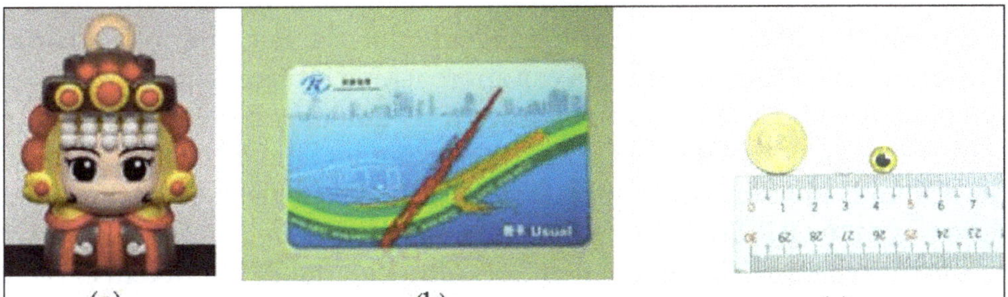

(a) (b) (c)

The sample of RFID Tag; a.3D Toy, b. plastic card, c. button tag.

In addition to the RFID card, various styles of RFID tag may be required by different consumer. According to the environment, the RFID tag may be designed to anti-water or anti-wet. Some RFID tags need to be applied to the metal environment such as containers. Due to the characteristics of RFID, suitable frequency of RFID corresponding to the environment should be selected carefully. In other words, the cost of different RFID may be also different according to the applications and environment. Furthermore, different shapes of RFID tag also costs. For example, the cost of an RFID tag embedded into a 3D toy is much more expensive than that embedded in a plastic card.

Hardware and Integration

The RFID Antenna is the main component for RFID tag induction. The antenna continuously spreads the electromagnetic wave. The energy is transmitted to the RFID tag. After induction, the RFID Antenna also receives the signal from the RFID tag.

After receiving the signal, the RFID Reader translates the signal into the digital data such as the UID of this RFID tag. Then, the RFID Reader sends the digital data to the corresponding systems or applications. To implement the RFID system, not only the RFID tag but also RFID reader and antenna hardware should be considered. Due to the design and product limitation, the RFID antenna cannot dynamically change. Therefore, similar to the RFID tag, the environment of the RFID system affects the size and cost of the antenna. The size of the antenna increases, the costs also increases. In addition, considering the implementation environment, to place the antenna at the suitable location for signal receiving also affects the performance and costs. Therefore, to integrate the RFID system with the existed system, some additional problems may need to be overcome.

The existed applications or systems should include or integrate the RFID system. In other words, to integrate the RFID system, the original working practices of the existed system may be changed which needs extra costs. For example, to identify goods in warehouse, original applications or systems may only need the manual operation. However, to integrate the RFID system, some infrastructure such as the placement of RFID antenna, wire line for connection between antenna and reader, and the establishment of RFID reader and system server are required. In other words, the extra costs of RFID infrastructure are needed. In addition, some existed systems are based on mechanical operation without too much intelligent analysis. For example, a car parking lot only needs to open the gate when a car approaching or according to the teleswitch. When integrated with the RFID system, the RFID antenna should be placed in front of the gate for induction. All the cars to the parking lot should present the RFID tag given on demand. In addition, the RFID reader should be used to analyze the signal information from the RFID tag. The server which includes the database should be used to judge whether the gate should open or not. Although the RFID system enhances the automation with less manual operation, some extra costs and delay may also happen. Therefore, the benefits of RFID system integration such as automation, information exchanging with third party applications, etc., are very important. Only when the benefits or additional new functions overcome the extra costs of RFID system integration then the integration of system will be used.

Plug and Play Middleware

In the RFID systems or applications, there are two partitions: RFID devices (includes RFID tag, antenna, and reader) and other devices or systems. Therefore, the application or middleware for communicating these two parts is needed. When using the

RFID device, the third party systems or applications should obtain the information from RFID devices.

Due to that there are many types of RFID hardware, the application program interface (API) for the communication between RFID Hardware and different third party applications is needed. In addition, the end user's devices are also various. Hence, the plug and play middleware for different hardware and applications is important.

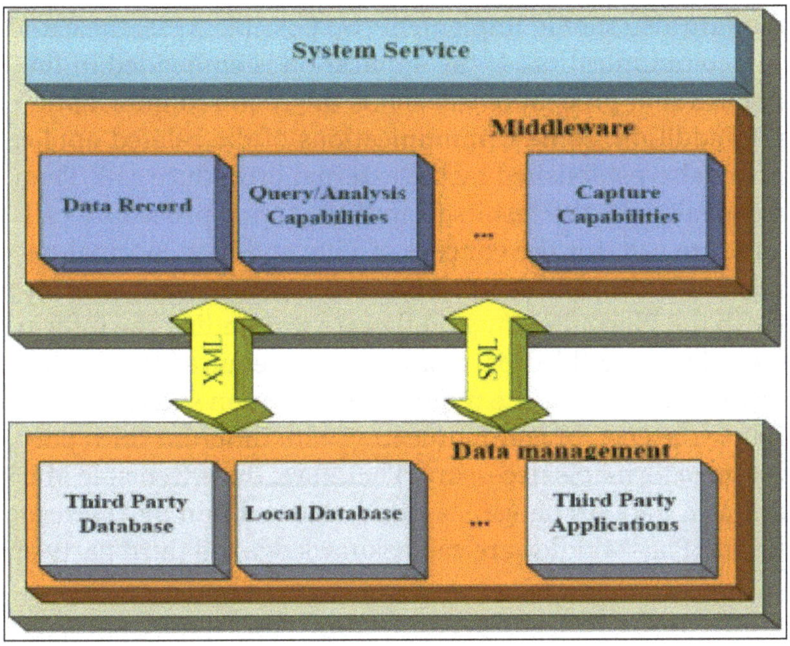

To manage the RFID information from different RFID Hardware, and the communication with different applications, the Plug and Play Middleware is proposed. To realize the concept of Plug and Play, the proposed middleware has to manage the information from the all possible third party RFID Hardware, deal with and parse the information, and then provide the required information to the corresponding applications. Therefore, the main purposes of the proposed Plug and Play Middleware are:

To parse the information from the RFID Hardware. Due to that there are different RFID product, the RFID parser is needed for analyzing and parsing the information from RFID Hardware. The information about UID, password, etc. will be parsed as the string for the further execution of applications. First, the Plug and Play Middleware provides the remote procedure call (RPC) function for the third party RFID Hardware. The UID of the RFID tag inducted by the RFID Hardware will be formulated as the string. In addition, the password or requirements for further information such as decryption code recorded in the End User RFID Device can be provided by the remote procedure call function. Second, for general communication, the Plug and Play Middleware also provides the sever-client socket link between the RFID Hardware and the middleware. In other words, even the RFID Hardware cannot implement the remote procedure call, depends on sever-client socket link, the information can be transmitted between Plug and Play Middleware and RFID Hardware.

To provide the application program interface (API). Since the RFID Hardware may not directly communicate with the applications, the Plug and Play Middleware has to implement the corresponding API for other third party applications or software.

The structure chart of data exchange in Database management. Furthermore, the Plug and Play Middleware also should implement two possible APIs: the external procedure call and network communication. If the application is embedded in the Plug and Play Middleware, the external procedure call sends the required information to the specific application. In addition, some communications of the related applications such as database query are also established by the external procedure call. Then, the Plug and Play Middleware deals with the results from the external procedure call. In opposition to external procedure call, for the concept of Plug and Play, normal network communication should also be implemented. Most third party software or applications can communicate with the Plug and Play Middleware via sending the information in string format. For example, if the third party application requires the further checking, the Plug and Play Middleware sends the required information such as UID to the server via Internet. To reduce the cost for communicating with different third party applications, the unify data storage format is necessary. Therefore, the eXtensible Markup Language (XML) can used as a data exchange standard. After obtaining the response from the server, the Plug and Play Middleware can acknowledge the third party application.

5

Applications

Radio frequency identification has a wide-range of applications such as in logistics and supply chain visibility, item level inventory tracking, attendee tracking, materials management, access control, IT asset tracking, library systems, interactive marketing, etc. This chapter discusses these applications of radio frequency identification in detail.

REAL WORLD USES

Logistics and Supply Chain Visibility

Walmart.

Winning in the supply chain means increasing efficiency, reducing errors, and improving quality. In chaotic manufacturing, shipping, and distribution environments, real-time data on the status of individual items provides insights that turn into actionable measures. With the visibility provided by RFID, you'll be on your way to Six Sigma Master Black Belt status in no time.

Item Level Inventory Tracking

Tracking assets on the item level is beneficial across a broad cross-section of industries, but the retail sector has one of the highest ceilings in terms of opportunity from the use of RFID. Tracking items through the supply chain is wonderful, but now think about tracking items through the supply chain all the way to the point of sale. With a well designed inventory system sharing data across all business units, you'll have a treasure trove of actionable data. One last added benefit — store employees can count inventory in a matter of minutes with a handheld RFID reader.

Race Timing

Timing marathons and races are one of the most popular uses of RFID, but often race participants never realize they're being timed using RFID technology, and that's a testament to RFID's ability toprovide a seamless consumer experience.

Attendee Tracking

If you've ever managed a large conference before, you'll know that it's key to keep the flow of traffic moving at a steady pace, especially in and out of seminars. With an RFID attendee solution, eliminate the need for registration lines at entrances.

Materials Management

In construction and other related industries, materials are often the largest project expenditure. On large job sites, simply finding materials can be problematic. RFID solutions like Jovix take the guess work out of the equation.

Access Control

Certain areas require an expected level of security and access. From doors to parking lots, RFID access control tags restrict access to only those pre-approved.

IT Asset Tracking

IT assets such as server blades, laptops, tablets, and other peripherals are costly investments for any company, not to mention that information stored on those items could prove detrimental in the wrong hands. IT asset tags give your IT team the ability to quickly do an inventory count and make sure everything is in place.

Tool Tracking

For industries that rely on a large variety of tools, fasteners, and other items, managing the availability of those assets is a frustrating process. Depending on the level of complexity, RFID tool tracking systems track which tools have been grabbed, which employees have taken resources, and which resources haven't been returned to the tool crib.

Kiosks

The Campus Crop.

Many kiosks use RFID to either manage resources or interact with users. DVD rental kiosks use RFID DVD tags to make sure customers receive their selected movie rental. Other examples of RFID kiosks include interactive media displays where an embedded RFID reader interrogates badges or cards.

Library Systems

An RFID library solution improves the efficiency of circulation operations. While barcodes require line of sight, RFID tags can be read from multiple angles which means the checkout and check-in process is significantly faster. The retail section, taking inventory of books on the shelf is dramatically faster.

Laundry Management

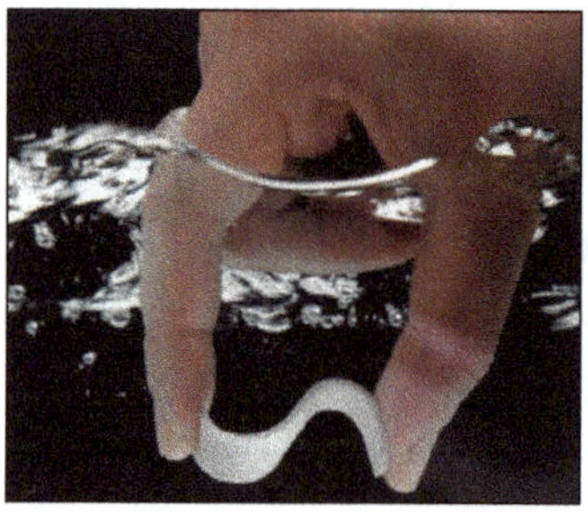

Large companies like casinos often manage thousands of employee uniforms. With an RFID laundry management system, operations can track which uniforms were assigned to specific employees, the age of uniforms, the number of times washed, and identify missing uniforms. RFID laundry tags provide a new level of visibility for laundry management.

Interactive Marketing

Example of Lexus' NFC Smart Poster in Wired Magazine.

RFID in marketing brings a certain level of interaction to campaigns. Whereas traditional advertising campaigns push a message onto the consumer, interactive campaigns invite the consumer to engage with the brand.

RTLS (Real Time Location System)

In some applications, you need to track the real-time location of assets, employees, or customers. Whether you're measuring the efficiency of worker movements, the effectiveness of a store floor plan, or tracking the location of valuable resources, RFID systems provide visibility in any number of locations.

RADIO FREQUENCY TECHNOLOGY IN SUPPLY CHAIN MANAGEMENT

If technology continues to advance, humans may be obsolete in the supply chain as a result of both automation and RFID utilization. Both technologies have been introduced into the supply chain over the past decade, resulting in a major shift in performance and costs. While automation is a process which can navigate product throughout facilities, RFID can improve efficiency dramatically as well as being the major driver for the elimination of human involvement in the supply chain.

RFID (Radio Frequency Identification) is a form of extremely low-power data communication between a RFID scanner and an RFID tag. The tags are placed on any number of items, ranging from individual parts to shipping labels. The RFID tag itself consists of a microchip and antennae, usually without a battery to power it. The tags can be printed using special printers, which wirelessly load the identifying information to the tags. The information on the tags can be used for a wide variety of tasks. When an item goes through the RFID scanners, information is read from the tag, which could include any amount of information, such as:

- Order ID number.
- Product bin location.
- Order status.
- Serial numbers for individual product components.
- Location logs.

The information is not limited to just holding ID and serial numbers. Since the information can be updated and transferred through any RFID receiver when in range, it can be joined with other software to update databases, send information online, and more. The amount of information that RFID can provide can be matched with order management systems to track shipment and stock locations automatically as the products move through warehouses and trucks. Powerful receivers can also track the exact location of products within a warehouse in real time, instead of relying on time and location logs to determine the location. This has been a major benefit to UPS, who've had RFID tracking chips in their warehouses for over ten years.

The largest problem facing supply chain management today is the potential for the occurrence of devastating errors. These errors could lead to trucks leaving product behind, not stocking to capacity, losing product, or delivering to the wrong locations. There are numerous reasons for such errors to occur, but the most common factor in their inception is human error. Much of the errors faced result from manual inspection and control over product flow in warehouses by humans. Many systems still operate using written data tables and checklists, which can be written incorrectly. Despite major advancements in technology over the past decade, the digital world is being sourced by analog supply chains. This is an aspect which RFID control can create beneficial change.

RFID can lead to completely autonomous warehouses and distribution centers. This can be seen in current warehouses with high levels of automation, where picking machines transport bins of product to a central sorting area to be boxed, before returning to their location. Zebra uses this technique to ensure product location accuracy as a means of reducing lost merchandise. Zebra places a great deal of importance on product visibility and transparency as well as accuracy in locations. Implementation of RFID into this system can ensure that both the correct product and the correct quantities of product are collected at both points, thereby eliminating errors seen in traditional analog supply networks. Coupled with the potential for self-driving trucks and the ever expanding internet of things in the cloud, product information can be tracked at all stages of shipment and storage, increasing accuracy, efficiency, and accountability. Where warehouses and receiving departments had to endure a certain amount of shrinkage of products in the past, a fully utilized RFID-enabled supply network can determine exactly where product is at all times, ensuring that theft is discovered immediately and enforced.

While the reduction of human interaction in the supply chain and warehouses in general can be eliminated with enough technology, it is uncertain to what degree this technology will be implemented in all aspects of supply chain management. While there are huge benefits to operating in such a system, the major negatives include high initial costs and restructuring costs associated with implementing this network into existing, analog networks. Humanity may not be completely removed from the supply chain in the future, but this technology, as well as more advanced techniques for automation, will increase the efficiency and reduce the costs required to operate these newer supply chains dramatically while significantly reducing errors and inefficiencies.

TRANSPONDER TIMING

Transponder timing (also called chip timing or RFID timing) is a technique for measuring performance in sport events. A transponder working on a radio-frequency identification (RFID) basis is attached to the athlete and emits a unique code that is detected by radio receivers located at the strategic points in an event.

Prior to the use of this technology, races were either timed by hand (with operators pressing a stopwatch) or using video camera systems.

A Jaguar Timing System at a finish line using RFID technology with overhead antennas and passive, disposable chips.

A ChronoTrack race controller with RFID antennas for detecting transponders attached to runner's shoes.

Transponder Systems

Generically, there are two types of transponder timing systems; active and passive. An active transponder consists of a battery-powered transceiver, connected to the athlete, that emits its unique code when it is interrogated.

A passive transponder does not contain a power source inside the transponder. Instead, the transponder captures electromagnetic energy produced by a nearby exciter and utilizes that energy to emit a unique code.

In both systems, an antenna is placed at the start, finish, and in some cases, intermediate time points and is connected to a decoder. This decoder identifies the unique transponder code and calculates the exact time when the transponder passes a timing point. Some implementations of timing systems require the use of a mat on the ground at the timing points while other systems implement the timing points with vertically oriented portals.

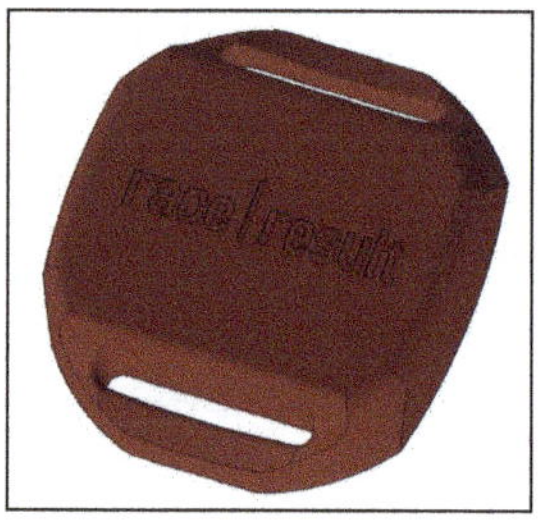

Active chip timing transponder.

ChampionChip.

Usage

Very large running events (more than 10,000) and triathlons were the first events to be transponder (or chip) timed because it is near impossible to manually time them. Also for large runs there are delays in participants reaching the start line, which penalize their performance. Some races place antennas or timing mats at both the start line and the finish line, which allow the exact net time to be calculated. Awards in a race are generally based on the "gun time" (which ignores any delay at the start) as per IAAF and USA Track and Field rules. However, some races use "net time" for presenting age group awards.

Disposable bib with two passive timing chips at the back.

In the past the transponder was almost always worn on the athletes running shoe, or on an ankle band. This enabled the transponder to be read best on antenna mats because the distance between the transponder and readers antenna is minimized offering the best capture rate. Transponders may be threaded onto the shoe laces for running. For triathlon a soft elastic ankle band holds the transponder to the leg and care is taken to ensure the transponder is in the correct orientation or polarity for maximum read performance. Transponders have also been placed on the race bib. In the past 5 years the newer UHF systems use transponders placed on the shoe lace, or stuck to the race number bib. In both cases, care must be taken to ensure the UHF tag does not directly touch a large part of the skin as this affects read performance. Despite this, UHF Systems have read performances as good (if not better) than the conventional low and high frequency systems. Because these UHF tags are made in huge volumes for industrial applications, their price is much lower than that of conventional re-usable transponders and the race does not bother to collect them afterwards. As of 2015, many UHF timers use a combination of ground antennas with panel antenna(s) mounted on a tripod at the side of the race course.

Back side of disposable RFID tag used for race timing.

All RFID timing systems incorporate a box housing the reader(s) with peripherals like a microprocessor, serial or ethernet communications and power source (battery). The readers are attached to one or more antennas that are designed for the particular operating frequency. In the case of low or medium frequencies these consist of wire loops incorporated into mats that cover the entire width of the timing point. For UHF systems the antennas consist of patch antennas that are protected in a matting system. The patch antennas may also be placed on stands or a finish gantry pointing towards the oncoming athlete. In most cases the distance between reader and antennas is restricted. Also more equipment is needed for events that require multiple timing points. Wider timing points require more readers and antennas. For active systems a simple wire loop is all that is needed since the transponder has its own power source and the loop serves as a trigger to turn on the transponder, then receive the relatively strong signal from the transponder. Therefore, active systems need less readers (or decoders) per timing point width.

All systems utilize specialized software to calculate results and splits. This software usually resides on a separate PC computer that is connected to the readers via serial or ethernet communications. The software relates the raw transponder code and timestamp data to each entrant in a database and calculates gun and net times of runners, or the splits of a triathlete. In advanced systems these results are instantly calculated and published to the internet so that athletes and spectators have access to results via any web enabled device.

ITEM-LEVEL TAGGING

Item-level tagging (or RFID item-level tagging, also known as ILT) is the tagging of individual products, as opposed to case-level and pallet-level tagging. Item-level tagging is used to track individual items in order to better control inventory, by providing retailers with the ability to tag individual items on the retail floor. Previously, RFID tags were used to track pallets of merchandise, rather than individual items, through the supply chain. With the use of printed RFID tags, retailers are now able to track inventory at the item level, scan the tag, and know the location.

Retailers are pushing for tagging each individual item. In fact, large companies like Wal-Mart, JC Penney, and Dillard's are issuing electronic product code mandates, where they request their suppliers to comply with these EPC protocols. In 2005, it was required that the suppliers use RFID tagging at the pallet and case level, but now it is required that they tag on the item-level as well. The reason why is it so important for them to implement this is because they want to avoid losing a sale over an out-of-stock item, which they believe accounts for a big part of their losses. Also, if they know where an item is at all times then it easier to move it to where it is supposed to be. By doing this they reduce transportation costs, they gain added shelf visibility and it drives down wasteful overstock.

Benefits

Item-level tagging provides a quick, automated, cost efficient and accurate way to track inventory through the supply chain and in the retail environment. Benefits to item-level tagging include better visibility and control of inventory and an expansion of customer experience capabilities. Item-level tagging is critical in order to determine how much inventory is on the floor, what sizes and colors need to be restocked and what inventory is available in stock rooms. Other benefits include the ability to keep a fully stocked floor, increased time and labor savings, increase inventory accuracy, and reduction in clearance items due to incorrect inventory and excess ordering.

Industries using Item-level Tagging

- Apparel.
- Consumer goods.
- Electronic goods.

ELECTRONIC ARTICLE SURVEILLANCE

Electronic article surveillance is a technological method for preventing shoplifting from retail stores, pilferage of books from libraries or removal of properties from office buildings. Special tags are fixed to merchandise or books. These tags are removed or deactivated by the clerks when the item is properly bought or checked out. At the exits of the store, a detection system sounds an alarm or otherwise alerts the staff when it senses active tags. Some stores also have detection systems at the entrance to the restrooms that sound an alarm if someone tries to take unpaid merchandise with them into the restroom. For high-value goods that are to be manipulated by the patrons, wired alarm clips called spider wrap may be used instead of tags.

Types

There are several major types of electronic article surveillance systems:

- Electro-Magnetic, also known as magneto-harmonic or Barkhausen effect.
- Acousto-magnetic, also known as magnetostrictive.
- Radio Frequency (8.2 MHz).
- Microwave.
- Video surveillance systems (to some extent).
- Concealed EAS Surveillance Systems.

Concealed EAS Systems

Callidus Concealed Solution.

These new trend of systems have caught on lately as their are no visible pedestals or hindrance in the store facade- but these are systems installed below the floor and dropped from the ceiling these can then protect merchandise of retailers from being stolen as the entire range of door is covered. Various studies and researches have been done into making this technology powerful and effective. There are site conditions and other parameters which enable them to be successfully installed but often it has now been noted that malls insist on concealed system as a mandate to increase the shopping experience- for example a part of Dubai Mall in United Arab Emirates do not allow any visible systems. Concealed Systems will be the way to go for the future.

Electro-magnetic Systems

EM labels are available in different sizes and colors.

These tags are made of a strip of amorphous metal (metglas) which has a very low magnetic saturation value. Except for permanent tags, this strip is also lined with a strip of ferromagnetic material with a moderate coercive field (magnetic "hardness"). Detection is achieved by sensing harmonics and sum or difference signals generated by the non-linear magnetic response of the material under a mixture of low-frequency (in the 10 Hz to 1000 Hz range) magnetic fields.

When the ferromagnetic material is magnetized, it biases the amorphous metal strip into saturation, where it no longer produces harmonics. Deactivation of these tags is therefore done with magnetization. Activation requires demagnetization.

The EM systems are suitable for libraries to protect books and media. In the retail segment, unlike AM and RF, EM can protect small or round items and products with foil packaging or metal objects, like cosmetics, baby milk cans, medicines, DIY tools, homeware etc. EM systems can also detect objects placed in foil bags or in metal briefcases.

A further application is the Intellectual property (IP) protection against theft: Security paper with embedded microwires, which is used to detect confidential documents if they are removed from a building.

Acousto-magnetic Systems

These are similar to magnetic tags in that they are made of two strips: a strip of magnetostrictive, ferromagnetic amorphous metal and a strip of a magnetically semi-hard metallic strip, which is used as a biasing magnet (to increase signal strength) and to allow deactivation. These strips are not bound together but free to oscillate mechanically.

A cutaway image of an acousto-magnetic tag.

Amorphous metals are used in such systems due to their good *magnetoelastic coupling*, which implies that they can efficiently convert magnetic energy into mechanical vibrations.

The detectors for such tags emit periodic tonal bursts at about 58 kHz, the same as the resonance frequency of the amorphous strips. This causes the strip to vibrate longitudinally by magnetostriction, and it continues to oscillate after the burst is over. The vibration causes a change in magnetization in the amorphous strip, which induces an AC voltage in the receiver antenna. If this signal meets the required parameters (correct frequency, repetition, etc.), the alarm is activated.

When the semi-hard magnet is magnetized, the tag is activated. The magnetized strip makes the amorphous strip respond much more strongly to the detectors, because the DC magnetic field given off by the strip offsets the magnetic anisotropy within the amorphous metal. The tag can also be deactivated by demagnetizing the strip, making the response small enough so that it will not be detected by the detectors.

AM tags are three dimensional plastic tags, much thicker than electro-magnetic strips and are thus seldom used for books. However, they are relatively inexpensive and have better detection rates (fewer false positives and false negatives) than electro-magnetic tags.

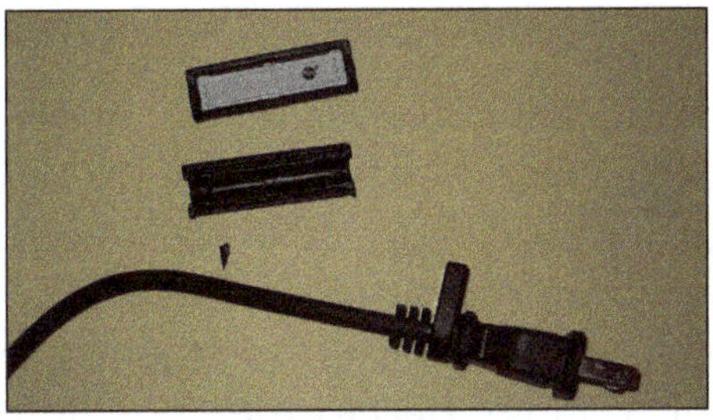

Emtag-removed.

Called Emtag by B&G International, this type of tag is often attached to the inside of a plastic surround permanently attached to the power cords of hand tools and equipment.

Radio Frequency Systems

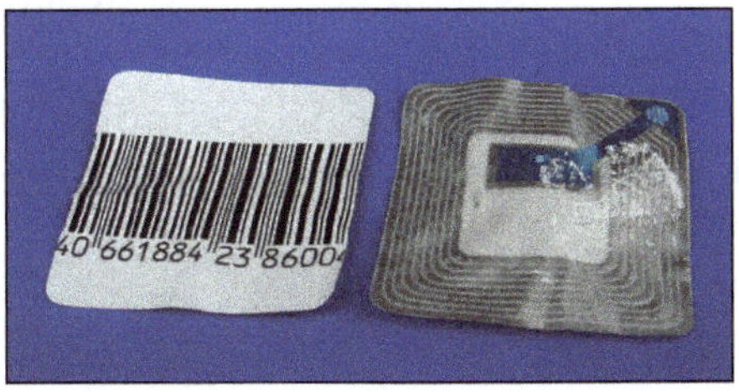

RF label.

RF tag placed under the label of a liquor bottle.

Series 304 RF EAS label.

These tags are essentially an LC tank circuit (L for inductor, C for capacitator) that has a resonance peak anywhere from 1.75 MHz to 9.5 MHz. The standard frequency for retail use is 8.2 MHz. Sensing is achieved by sweeping around the resonant frequency and detecting the dip.

Deactivation for 8.2 MHz label tags is typically achieved using a deactivation pad. In the absence of such a device, labels can be rendered inactive by punching a hole, or by covering the circuit with a metallic label, a "detuner". The deactivation pad functions by partially destroying the capacitor. Though this sounds violent, in reality, both the process and the result are unnoticeable to the naked eye. The deactivator causes a micro short circuit in the label. This is done by submitting the tag to a strong electromagnetic field at the resonant frequency, which induces voltages exceeding the capacitor's breakdown voltage.

In terms of deactivation, Radio Frequency is the most efficient of the 3 technologies (RF, EM, AM – there are no microwave labels) given that the reliable "remote" deactivation distance can be up to 30 cm (11.8 in). It also benefits the user in terms of running costs, since the RF de-activator only activates to send a pulse when a circuit is present. Both EM and AM deactivation units are on all the time and consume considerably more electricity. The reliability of "remote" deactivation (i.e. non-contact or non-proximity deactivation) capability makes for a fast and efficient throughput at the checkout.

Efficiency is an important factor when choosing an overall EAS solution given that time lost attempting to deactivate labels can be an important drag of cashier productivity as well as customer satisfaction if unwanted alarms are caused by tags that have not been effectively deactivated at the point of sale.

Deactivation of RF labels is also dependent on the size of the label and the power of the deactivation pad (the larger the label, the greater the field it generates for deactivation to take place. For this reason very small labels can cause issues for consistent deactivation). It is common to find RF deactivation built into barcode flat and vertical scanners at the POS in food retail especially in Europe and Asia where RF EAS technology has been the standard for nearly a decade. In apparel retail deactivation usually takes the form of flat pads of approx. 30x30 cm.

Microwave Systems

These permanent tags are made of a non-linear element (a diode) coupled to one microwave and one electrostatic antenna. At the exit, one antenna emits a low-frequency (about 100 kHz) field, and another one emits a microwave field. The tag acts as a mixer re-emitting a combination of signals from both fields. This modulated signal triggers the alarm. These tags are permanent and somewhat costly. They are mostly used in clothing stores and have practically been withdrawn from use.

Source Tagging

Source tagging is the application of EAS security tags at the source, the supplier or manufacturer, instead of at the retail side of the chain. For the retailer, source tagging eliminates the labor expense needed to apply the EAS tags themselves, and reduces the time between receipt of merchandise and when the merchandise is ready for sale. For the supplier, the main benefit is the preservation of the retail packaging aesthetics by easing the application of security tags within product packaging. Source tagging allows the EAS tags to be concealed and more difficult to remove.

The high-speed application of EAS labels, suited for commercial packaging processes, was perfected via modifications to standard pressure-sensitive label applicators and was developed and introduced by Craig Patterson, initially for Hewlett Packard print cartridges. Today, consumer goods are source tagged at high speeds with the EAS label incorporated into the packaging or the product itself.

The most common source tags are AM strips and 8.2 MHz radio frequency labels. Most manufacturers use both when source tagging in the USA. In Europe there is little demand for AM tagging given that the Food and Department Store environments are dominated by RF technology.

Tag Pollution

One significant problem from source tagging is something called "tag pollution" caused when non-deactivated tags carried around by customers cause unwanted alarms, decreasing the effectiveness and integrity of the EAS system. The problem is that no store has more than one system. Therefore, if a store actually has an anti-shoplifting system to deactivate a label they will only deactivate the one that is part of their system. If a store does not use an EAS system, they will not deactivate any tags at all. This is often the reason why people trigger an alarm entering a store, which can cause great frustration for both customers and staff. The problem is most evident in shopping malls where customers wander between stores.

Occasional vs. Professional Shoplifters

EAS systems can provide a solid deterrent against casual theft. The occasional shoplifter, not being familiar with these systems and their mode of operation, will either get caught by them, or preferably, will be dissuaded from attempting any theft in the first place.

Informed shoplifters are conscious of how tags can be removed or deactivated. A common method of defeating RF tags is the use of so-called booster bags. These are typically large paper bags that have been lined with multiple layers of aluminium foil to effectively shield the RF label from detection, much like a Faraday cage. A similar situation would be the loss of signal that a cell phone suffers inside an elevator: The electro-magnetic, or radio, waves are effectively blocked, reducing the ability to send or receive information.

However, they may miss some tags or be unable to remove or deactivate all of them, especially if concealed or integrated tags are used. As a service to retailers, many manufacturers integrate security tags in the packaging of their products, or even inside the product itself, though this is rare and not especially desirable either for the retailer or the manufacturer. The practical totality of EAS labels are discarded with the product packaging. This is of particular application in everyday items that consumers might carry on their person to avoid the inconvenience of potentially live reactivated EAS tags when walking in and out of retail stores.

Hard tags, typically used for clothing or ink tags, known as benefit denial tags, may reduce the rate of tag manipulation. Also, shoplifters deactivating or detaching tags may be spotted by the shop staff.

Shoplifting tools are illegal in many jurisdictions, and can, in any case, serve as evidence against the perpetrators. Hence, informed shoplifters, although they decrease their risk of being caught by the EAS, expose themselves to much greater judicial risks if they get caught with tools, booster bags, or while trying to remove tags, as this shows intent to steal.

The possession of shoplifting tools (e.g. lined bags or wire cutters to cut bottle tags) can lead to the suspect being arrested for suspicion of theft or "Going equipped for stealing, etc." within the UK judicial system.

In summary, while even the least expensive EAS systems will catch most occasional shoplifters, a broader range of measures are still required for an effective response that can protect profits without impeding sales.

Tags Containing Alarms

Tags can be equipped with a built-in alarm which sounds when the tag detects tampering or unauthorized removal from the store. The tag not only triggers the store's electronic article surveillance system, but also sounds an alarm attached to the merchandise. The local alarm continues to sound for several minutes after leaving the store, attracting attention to the shopper carrying the merchandise.

Installation Costs

A single EAS detector, suitable for a small shop, is accessible to all retail stores, and should form a part of any coherent loss or profit protection system.

Disposable tags cost a matter of cents and may have been embedded during manufacture. More sophisticated systems are available, which are more difficult to circumvent.

These solutions tend to be product category specific as in the case of high value added electronics and consumables; consequently they are more expensive. Examples are "Safers", transparent secure boxes that completely enclose the topic to be protected, Spiders that wrap around packaging and Electronic Merchandise Security Systems that allow phones and tablets to be used securely in the store before purchase. All of these require specific detachers or electronic keys at the point-of-sale desk. They have the advantages of being reusable, strong visual deterrents to potential theft.

Tag Orientation

Except for microwave, the detection rate for all these tags depends on their orientation relative to the detection loops. For a pair of planar loops forming a Helmholtz coil, magnetic field lines will be approximately parallel in their center. Orienting the tag so that no magnetic flux from the coils crosses them will prevent detection, as the tag won't be coupled to the coils. This shortcoming, documented in the first EAS patents, can be solved by using multiple coils or by placing them in another arrangement such as a figure-of-eight. Sensitivity will still be orientation-dependent but detection will be possible at all orientations.

Detaching

A detacher is used to remove re-usable hard tags. The type of detacher used will depend on the type of tag. There are a variety of detachers available, with the majority using powerful magnets. Any store that uses an anti-shoplifting system and has a detacher should take care to keep it secured such that it cannot be removed. Some detachers actually have security tags inside them, to alert store personnel of them being removed from (or being brought into) the store. With increasing prevalence, stores have metal detectors at the entrance that can warn against the presence of booster bags or detachers.

Electro-magnetic Activation and Deactivation

Deactivation of magnetic tags is achieved by straightforward magnetization using a strong magnet. Magneto-acoustic tags require demagnetization. However, sticking a powerful magnet on them will bias disposable magnetic tags and prevent resonance in magneto-acoustic tags. Similarly, sticking a piece of metal, such as a large coin on a disposable radio-frequency tag will shield it. Non-disposable tags require stronger magnets or pieces of metal to disable or shield since the strips are inside the casing and thus further away. Deactivation of some EAS tags can trigger signals sent to the cashier alerting them.

Shielding

Most systems can be circumvented by placing the tagged goods in a bag lined with aluminum foil. The booster bag will act as a Faraday cage, shielding the tags from the antennas. Although some vendors claim that their acousto-magnetic systems cannot be defeated by bags shielded with aluminum foil, a sufficient amount of shielding (in the order of 30 layers of standard 20 μm foil) will defeat all standard systems.

Although the amount of shielding required depends on the system, its sensitivity, and the distance and orientation of the tags relative to its antennas, total enclosure of tags is not strictly necessary. Indeed, some shoplifters use clothes lined with aluminum foil. Low-frequency magnetic systems will require more shielding than radio-frequency systems due to their use of near-field magnetic coupling. Magnetic shielding, with steel or mu-metal, would be more effective, but also cumbersome and expensive.

The shielding technique is well known amongst shoplifters and store owners. Some countries have specific laws against it. In any case, possession of such a bag demonstrates a prior-intent to commit a crime, which in many jurisdictions raises shoplifting from misdemeanor to felony status, because they are considered a "burglary tool."

To deter the use of booster bags, some stores have add-on metal detector systems which sense metallic surfaces.

Jamming

Like most systems that rely on transmission of electromagnetic signals through a hostile medium, EAS sensors can be rendered inoperative by jamming. As the signals from tags are very low-power (their cross-section is small, and the exits are wide), jamming requires little power. Evidently, shoplifters will not feel the need to follow radio transmission regulations; hence crude, easy-to-build transmitters will be adequate for them. However, due to their high frequency of operation, building a jammer can be difficult for microwave circuits; these systems are therefore less likely to be jammed. Although jamming is easy to perform, it is also easy to detect. A simple firmware upgrade should be adequate for modern DSP-based EAS systems to detect jamming. Nevertheless, the vast majority of EAS systems do not currently detect it.

Patents

Radio-frequency systems have the most manufacturers because they are not covered by patents, use well-known technology derived from radio communications, use little power, and can be manufactured without expensive metal alloys. There are exceptions however for example; a patented die-cut circuit manufacturing process is held by Miyake Inc. of Japan. Also, patents have recently been appearing for various combination tag designs for integration of both RF and RFID. The last known patent is the "Global Guard" by Argos Global, an EAS system that integrates video surveillance hidden inside pedestals, getting a close up of potential thieves.

Retailers Advantage has become more prominent with their release of both the 'a3tag' and 'STPLUS' hard tags. They remain one the last developers who own patents on EAS and RFID hard tags.

Until recently, acousto-magnetic systems were covered by now-expired patents held by Sensormatic. WG Security Products, Inc. won a court battle against Sensormatic clarifying that WG Security Products acousto-magnetic systems did not infringe any

Sensormatic patents. Disposable acousto-magnetic tags require special metal alloys; non-disposable ones require more expensive ferrite cores.

Effects on Theft

A 2017 review study found that "based on the available evidence it is difficult to determine the effectiveness of tags as a theft reduction measure, albeit there is suggestive evidence that more visible tags are associated with greater reductions in theft than less visible tags".

MAGNETIC STRIPE CARD

An example of the reverse side of a typical credit card: Green circle #1 labels the Magnetic stripe Visualization of magnetically stored information on a magnetic stripe card.

A magnetic stripe card is a type of card capable of storing data by modifying the magnetism of tiny iron-based magnetic particles on a band of magnetic material on the card. The magnetic stripe, sometimes called swipe card or magstripe, is read by swiping past a magnetic reading head. Magnetic stripe cards are commonly used in credit cards, identity cards, and transportation tickets. They may also contain an RFID tag, a transponder device and a microchip mostly used for business premises access control or electronic payment.

Visualization of magnetically stored information on a magnetic stripe card.

Magnetic recording on steel tape and wire was invented in Denmark around 1900 for recording audio. In the 1950s, magnetic recording of digital computer data on plastic tape coated with iron oxide was invented. In 1960, IBM used the magnetic tape idea to develop a reliable way of securing magnetic stripes to plastic cards, under a contract with the US government for a security system. A number of International Organization for Standardization standards, ISO/IEC 7810, ISO/IEC 7811, ISO/IEC 7812, ISO/IEC 7813, ISO 8583, and ISO/IEC 4909, now define the physical properties of the card, including size, flexibility, location of the magstripe, magnetic characteristics, and data formats. They also provide the standards for financial cards, including the allocation of card number ranges to different card issuing institutions.

First Magnetic Striped Plastic Credit and Badge Access Cards

Front side of the first magnetic stripe plastic credit card. The narrow magnetic stripe is on the front of the card. It was later switched to the back side.

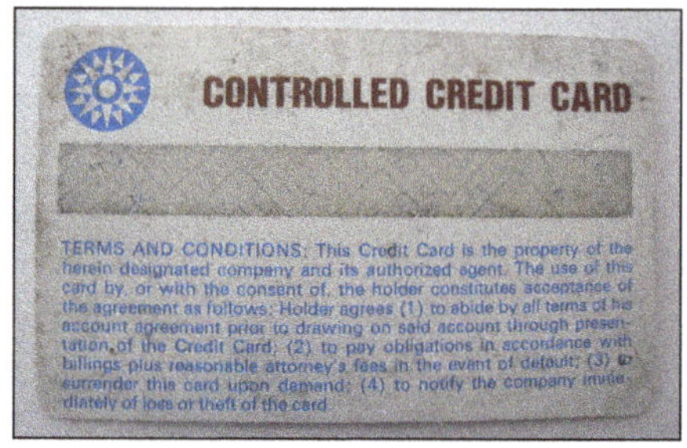

Back side of the first magnetic stripe plastic credit card.

The major development of the magnetic striped plastic card began in 1969 at the IBM Information Records Division (IRD) headquartered in Dayton N.J. In 1970, the marketing organization was transferred by IBM DPD back to the Information Records Division in order to begin sales and marketing strategies for the magnetically striped and encoded cards being developed. It took almost two years for IBM IRD engineers to not only develop the process for reliably applying the magnetic stripe to plastic cards via a hot stamping method, but also develop the process for encoding the magnetic stripe utilizing the IBM Delta Distance C Optical Bar Code format. This engineering effort resulted in IBM IRD producing the first magnetic striped plastic credit and ID cards used by banks, insurance companies, hospitals and many others. Another result of this project was that IBM IRD and IBM Data Processing Division announced on February 24, 1971 the first Magnetic Credit Card Service Center and the IBM 2730-1 Transaction Validation Terminal. Arthur E. Hahn Jr. was hired by IBM IRD in Dayton, N.J. on Aug 12, 1969 to head up this engineering effort. Other members of the group were David Morgan (Manager), Billy House (Software Developer), William Creeden (Programmer), and E. J. Gillen (Mechanical Engineering/Machining). They were given a recently announced IBM 360 Model 30 computer with 50k of RAM for control of the encoding/embossing of the Magnetic Stripe Cards. The IBM 360 computer was for scientific/business applications so the IRD engineers first had to convert the 360 into a "process control computer" and then develop software and hardware around it. Due to the limited RAM, the software was developed in 360 Assembler Language. This conversion enabled the 360 computer to monitor and control the entire production process the IRD engineers designed and built. The engineering design/build effort was carried out in a raised floor secured area of IBM IRD in Dayton, N.J. which was built specifically for the project. This tightly secured area with limited access was required because of the sensitivity of the data that would ultimately be used to encode and emboss the credit and ID cards.

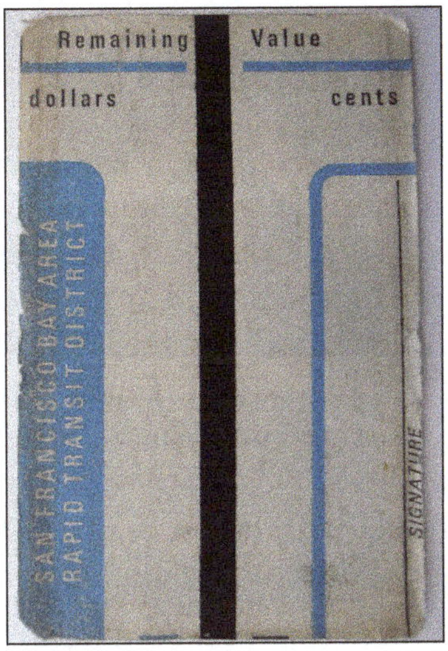

Back of early magnetic striped encoded paper card. The narrow magnetic stripe in the center of the card was applied using a magnetic slurry paint.

Bar Code Encoding Developments

The IRD engineers first had to develop a reliable process of hot stamping the magnetic stripe to the plastic cards. This was necessary in order to meet the close tolerances required to reliably encode and read the data on the Magnetic Stripe Cards by magnetic write/read heads. The magnetic stripe was encoded with a single track of data utilizing the IBM Delta Distance C Optical Bar Code format. The Delta Distance C Optical Bar Code was developed by the IBM Systems Development Division working at Research Triangle Park in Raleigh North Carolina headed up by George J. Laurer. Other members of the group were N. Joseph Woodland, Paul McEnroe, Dr. Robert Evans, Bernard Silver, Art Hamburgen, Heard Baumeister and Bill Crouse. The IBM group in Raleigh was competing with RCA, Litton-Zellweger and other companies who were working with the National Retail Merchants Association NRMA to develop a standard optical bar code to be used in the retail industry. NRMA wanted an optically readable code that could be printed on products allowing purchasers to rapidly "check out" at the new electronic cash register/checkout counters being developed. The code would also be used for production and inventory control of products. Of the many optical bar codes submitted to NRMA by IBM and other companies, NRMA finally selected the later version of the IBM bar code known as the Delta Distance D Optical Bar Code format. The Delta Distance C Code was an earlier version of the Universal Product Code (UPC). The UPC code was selected in 1973 by NRMA as their standard and has become the World Wide Standard that we all know today as the UPC Uniform Product Code.

Production

In 1971, after the IBM IRD engineers completed the development and building phase of the project they began in 1969, they released the equipment to the IRD manufacturing group in Dayton N.J. to begin producing the plastic magnetic striped credit and ID cards. Because of the sensitivity of the customer data and the security requirements of banks, insurance companies and others, the manufacturing group decided to leave the entire line in the secured area where it was developed.

Banks, insurance companies, hospitals etc., supplied IBM IRD with “raw plastic cards” preprinted with their logos, contact information etc. They also supplied the data information which was to be encoded and embossed on the cards. This data was supplied to IRD on large 0.5 inch wide, 10.5 inch diameter IBM Magnetic Tape Reels which was the standard for computers at that time. The manufacturing process started by first applying the magnetic stripe to the preprinted plastic cards via the hot stamping process developed by the IBM IRD engineers. This operation of applying the magnetic stripe to the plastic cards was done off line in another area of IBM IRD and not in the secured area. The cards were then brought into the secured area and placed in “hoppers” at the beginning of the production line. The tape reels containing the data were then installed on the modified IBM 360 computer prior to beginning the encoding, embossing and verification of the cards. After the 360 performed a check to verify that all systems and stations were loaded and ready to go, the computer began feeding the Magnetic Striped Plastic Cards from the hoppers at the front end of the production line down a motorized track. The entire operation was fully automated and controlled by the modified IBM 360 business computer. The line consisted of the following stations and operations:

- Plastic card feeder station: The cards were fed down a track in single file from card hoppers.
- Magnetic write/read encoding station: The IBM 360 computer sent over the data which was encoded on the magnetic stripe utilizing the IBM Delta Distance C Optical Bar Code format. The card passed under the read head and the encoded data was sent back to the 360 for verification.
- An embossing station: The IRD engineers purchased and modified a Data Card Corp embossing machine and interfaced it with the IBM 360 computer to emboss the cards. The original design concept called for an Addressograph-Multigraph embossing machine, however, the IRD engineers quickly switched to a Data Card Corp embossing machine. Data Card Corp, a Minneapolis/St. Paul company, had just developed the first electronically controlled embossing machine for plastic cards and effectively obsoleted all other mechanical operated embossers.
- A topping station: To highlight the embossing.
- An imprinter station: To imprint the embossing on an automatically fed paper roll.

- An optical reader station: To read the embossed information off the paper roll and feed it back to the 360 computer for verification.
- A one card rejection station: If either the encoding or embossing data on the card was not verified by the 360 computer, that one card was rejected. If both the encoded and embossed data was confirmed by the 360 computer, the card proceeded down the line.
- A mailer station: A mailer was printed with the name and address of the card holder along with the date and other relevant card information. These mailers were also preprinted and die cut by IRD according to the customers specs and logo requirements and were fed into the line out of boxes in a continuous fan feed method.
- A card insertion station: Here the card was automatically inserted onto the mailer.
- A bursting and folding station: Here the mailers were burst apart and then folded into a 3 fold packet that would fit into a business size envelope.
- An envelope printer/insertion station: Here an envelope was printed with the name and address of the customer and the mailer containing the card was automatically inserted into the envelope and sealed.

This completed the manufacturing line for the magnetic striped encoded and embossed plastic credit and badge access cards. The envelopes were then taken to be posted and mailed directly to the customers of the companies who had ordered the cards from IRD.

What this small engineering group at IBM IRD and the IBM Bar Code development group in Raleigh accomplished in developing the first magnetic stripe credit and ID cards cannot be overstated. They laid the foundation for the entire magnetic stripe card industry that we know and use today through our use of credit cards, ATM cards, ID cards, hotel room and access cards, transportation tickets, and all the terminals and card readers that read the cards and enter the data into computers. Their developments resulted in every person having the ability to easily carry a card that connects them directly to computers with all the ramifications thereof.

Neither IBM nor anyone else applied for or received any patents pertaining to the magnetic stripe card, the delta-distance barcodes or even the Uniform Product Code (UPC). IBM felt that with an open architecture, it would enhance the growth of the media thereby resulting in more IBM computers and associated hardware being sold. As with all new technologies, the magnetic stripe card developed and produced by IBM IRD with one track of encoded data using the Delta Distance C Bar Code format was quickly obsolete. Because of the electronic ATM/reservation/check out/and access systems that were rapidly developing, the banks, airlines and other industries required more encoded data. A wider magnetic stripe enabling multiple tracks of encoding along with new encoding standards was required. The first US Patents for the ATM were granted in 1972 and 1973.

Other groups within IBM and other companies continued on with expanding the work done by this small group of engineers at IBM IRD, however, the contributions that these IBM IRD engineers made to the development of the magnetic stripe card is analogous to the Wright Brothers' contribution to the airline industry of today.

Further Developments and Encoding Standards

There were a number of steps required to convert the magnetic striped media into an industry acceptable device. These steps included:

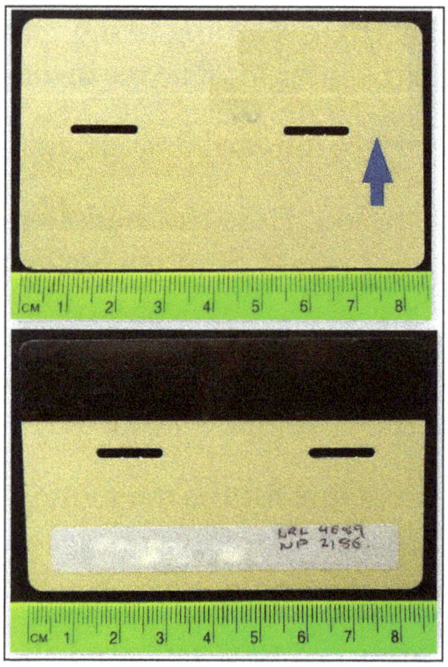

Example of a card from the late 1980s used in food vending machines in the UK.

- Creating the international standards for stripe record content, including which information, in what format, and using which defining codes.
- Field testing the proposed device and standards for market acceptance.
- Developing the manufacturing steps needed to mass-produce the large number of cards required.
- Adding stripe issue and acceptance capabilities to available equipment.

These steps were initially managed by Jerome Svigals of the Advanced Systems Division of IBM, Los Gatos, California from 1966 to 1975.

In most magnetic stripe cards, the magnetic stripe is contained in a plastic-like film. The magnetic stripe is located 0.223 inches (5.66 mm) from the edge of the card, and is 0.375 inches (9.52 mm) wide. The magnetic stripe contains three tracks, each 0.110 inches (2.79 mm) wide. Tracks one and three are typically recorded at 210 bits per inch (8.27 bits per mm), while track two typically has a recording density of 75 bits per inch (2.95 bits per mm). Each track can either contain 7-bit alphanumeric characters, or 5-bit numeric characters. Track 1 standards were created by the airlines industry (IATA). Track 2 standards were created by the banking industry (ABA). Track 3 standards were created by the thrift-savings industry.

Magstripes following these specifications can typically be read by most point-of-sale hardware, which are simply general-purpose computers that can be programmed to perform specific tasks. Examples of cards adhering to these standards include ATM cards, bank cards (credit and debit cards including Visa and MasterCard), gift cards, loyalty cards, driver's licenses, telephone cards, membership cards, electronic benefit transfer cards (e.g. food stamps), and nearly any application in which value or secure information is *not* stored on the card itself. Many video game and amusement centers now use debit card systems based on magnetic stripe cards.

Magnetic stripe cloning can be detected by the implementation of magnetic card reader heads and firmware that can read a signature of magnetic noise permanently embedded in all magnetic stripes during the card production process. This signature can be used in conjunction with common two-factor authentication schemes utilized in ATM, debit/retail point-of-sale and prepaid card applications.

Counterexamples of cards which intentionally ignore ISO standards include hotel key cards, most subway and bus cards, and some national prepaid calling cards (such as for the country of Cyprus) in which the balance is stored and maintained directly on the stripe and not retrieved from a remote database.

Magnetic Stripe Coercivity

Detailed visualization of magnetically stored information on a magnetic stripe card.

Magstripes come in two main varieties: high-coercivity (HiCo) at 4000 Oe and low-coercivity (LoCo) at 300 Oe, but it is not infrequent to have intermediate values at 2750 Oe. High-coercivity magstripes require higher amount of magnetic energy to encode, and therefore are harder to erase. HiCo stripes are appropriate for cards that are frequently used, such as a credit card. Other card uses include time and attendance tracking, access control, library cards, employee ID cards and gift cards. Low-coercivity magstripes require a lower amount of magnetic energy to record, and hence the card writers are much cheaper than machines which are capable of recording high-coercivity magstripes. However, LoCo cards are much easier to erase and have a shorter lifespan. Typical LoCo applications include hotel room keys, time and attendance tracking, bus/transit tickets and season passes for theme parks. A card reader can read either type of magstripe, and a high-coercivity card writer may write both high and low-coercivity cards (most have two settings, but writing a LoCo card in HiCo may sometimes work), while a low-coercivity card writer may write only low-coercivity cards.

In practical terms, usually low coercivity magnetic stripes are a light brown color, and high coercivity stripes are nearly black; exceptions include a proprietary silver-colored formulation on transparent American Express cards. High coercivity stripes are resistant to damage from most magnets likely to be owned by consumers. Low coercivity stripes are easily damaged by even a brief contact with a magnetic purse strap or fastener. Because of this, virtually all bank cards today are encoded on high coercivity stripes despite a slightly higher per-unit cost.

Magnetic stripe cards are used in very high volumes in the mass transit sector, replacing paper based tickets with either a directly applied magnetic slurry or hot foil stripe. Slurry applied stripes are generally less expensive to produce and are less resilient but are suitable for cards meant to be disposed after a few uses.

Financial Cards

There are up to three tracks on magnetic cards known as tracks 1, 2, and 3. Track 3 is virtually unused by the major worldwide networks, and often isn't even physically present on the card by virtue of a narrower magnetic stripe. Point-of-sale card readers almost always read track 1, or track 2, and sometimes both, in case one track is unreadable. The minimum cardholder account information needed to complete a transaction is present on both tracks. Track 1 has a higher bit density (210 bits per inch vs. 75), is the only track that may contain alphabetic text, and hence is the only track that contains the cardholder's name.

Track 1 is written with code known as DEC SIXBIT plus odd parity. The information on track 1 on financial cards is contained in several formats: A, which is reserved for proprietary use of the card issuer, B, which is described below, C-M, which are reserved for use by ANSI Subcommittee X3B10 and N-Z, which are available for use by individual card issuers:

Track 1

Format B

- Start sentinel: One character (generally '%').
- Format code="B": One character (alpha only).
- Primary account number (PAN): Up to 19 characters. Usually, but not always, matches the credit card number printed on the front of the card.

- Field Separator: One character (generally '^').
- Name: 2 to 26 characters.
- Field Separator: One character (generally '^').
- Expiration date: Four characters in the form YYMM.
- Service code: Three characters.
- Discretionary data: May include Pin Verification Key Indicator (PVKI, 1 character), PIN Verification Value (PVV, 4 characters), Card Verification Value or Card Verification Code (CVV or CVC, 3 characters).
- End sentinel: One character (generally '?').
- Longitudinal redundancy check (LRC): It is one character and a validity character calculated from other data on the track.

Track 2

This format was developed by the banking industry (ABA). This track is written with a 5-bit scheme (4 data bits + 1 parity), which allows for sixteen possible characters, which are the numbers 0-9, plus the six characters `: ; < = > ?`. The selection of six punctuation symbols may seem odd, but in fact the sixteen codes simply map to the ASCII range 0x30 through 0x3f, which defines ten digit characters plus those six symbols. The data format is as follows:

- Start sentinel: One character (generally ';').
- Primary account number (PAN): Up to 19 characters. Usually, but not always, matches the credit card number printed on the front of the card.
- Separator: One char (generally '=').
- Expiration date: Four characters in the form YYMM.
- Service code: Three digits. The first digit specifies the interchange rules, the second specifies authorization processing and the third specifies the range of services.
- Discretionary data: As in track one.
- End sentinel: One character (generally '?')
- Longitudinal redundancy check (LRC): It is one character and a validity character calculated from other data on the track. Most reader devices do not return this value when the card is swiped to the presentation layer, and use it only to verify the input internally to the reader.

Service code values common in financial cards:

First digit:

1: International interchange OK.

2: International interchange, use IC (chip) where feasible.

5: National interchange only except under bilateral agreement.

6: National interchange only except under bilateral agreement, use IC (chip) where feasible.

7: No interchange except under bilateral agreement (closed loop).

9: Test.

Second digit:

0: Normal.

2: Contact issuer via online means.

4: Contact issuer via online means except under bilateral agreement.

Third digit:

0: No restrictions, PIN required.

1: No restrictions.

2: Goods and services only (no cash).

3: ATM only, PIN required.

4: Cash only.

5: Goods and services only (no cash), PIN required.

6: No restrictions, use PIN where feasible.

7: Goods and services only (no cash), use PIN where feasible.

Other Card Types

Smart cards are a newer generation of card that contain an integrated circuit. Some smart cards have metal contacts to electrically connect the card to the reader, and contactless cards use a magnetic field or radio frequency (RFID) for proximity reading.

Hybrid smart cards include a magnetic stripe in addition to the chip — this is most commonly found in a payment card, so that the cards are also compatible with payment terminals that do not include a smart card reader.

Cards with all three features: magnetic stripe, smart card chip, and RFID chip are also becoming common as more activities require the use of such cards.

Vulnerabilities

DEF CON 24

During DEF CON 24, Weston Hecker presented *Hacking Hotel Keys, and Point Of Sales Systems*. In the talk, Hecker described the way magnetic strip cards function and utilised spoofing software, and an Arduino to obtain administrative access from hotel keys, via service staff walking past him. Hecker claims he used administrative keys from POS systems on other systems, effectively providing access to any system with a magnetic stripe reader, providing access to run privileged commands.

RFID IN COMMERCIAL CONSTRUCTION

RFID has a place in commercial construction because it provides the industry a potential to improve construction productivity, quality, safety, and economy, cutting labor and material costs and enhancing project schedules. Contractors are interested in discovering ways that RFID can save time to provide more efficient project management. However, do not look for RFID technologies to be implemented industry-wide in the near future. Current risks and challenges associated with developing RFID technologies pose obstacles to strong growth and implementation of RFID.

Industry fragmentation, slow technology deployment, complex supply chains, high cost of individual research and development, cost of overall RFID packages, and proprietor business concerns are the leading factors affecting the implementation of RFID applications in the field of construction. However, several companies and development groups are using pilot studies to identify viable construction applications in determining the economic potential and risks associated with emerging technologies.

The construction industry as a whole must collaborate to fully realize the potential for new technologies. The industry should feel obligated to freely communicate and publicize the results of their studies in RFID to lower the burden of cost on individuals, to establish an industry standard, and to minimize risks.

Construction Component and Materials Management

Tracking Materials

Future materials tracking management systems may be able to provide owners the ability to determine construction progress and materials delivered by simply walking a site where all materials are identified and tagged using an RFID system. This would guarantee more accurate estimates of percent construction complete and the number and quantity of delivered goods. The following case studies present the benefits of using an RFID tracking system to identify materials and deliverables. However, until RFID is cheap enough to implement across all construction applications, bar codes and RFID may be used simultaneously as shown in figure.

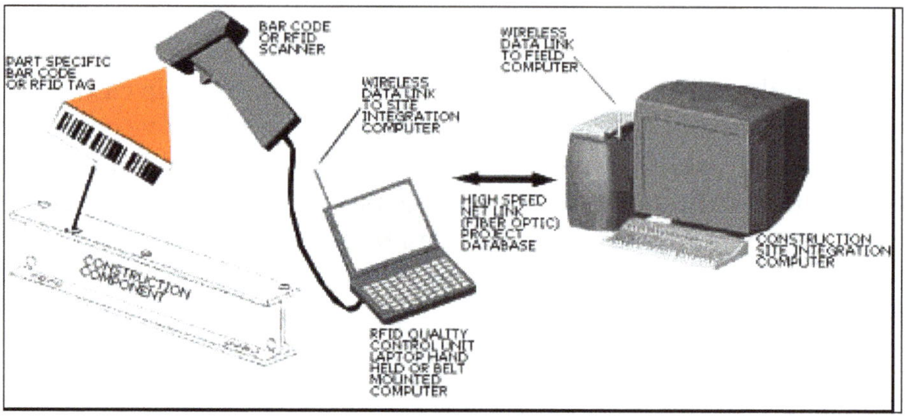

Schematic information flow for discrete component tracking.

Tracking Materials (Construction Industry Institute Pilot Studies)

The most conventional use RFID has seen thus far in construction is in its ability to improve the efficiency of the materials management process. Large amounts of time and money can be saved when materials can be properly identified and located. In fact, in a case study conducted by Bechtel in their $338 million Red Hills Project, time spent locating and tracking pipe supports and hangers was reduced by 30% (159 minutes per 100 hangers) as a result of the RFID system used. Also, the inspection process was much easier and more accurate than that of the traditional manual process.

Bechtel's Baytown Project was also a pilot study for the tracking of pipe supports on a coking unit at the Exxon Mobil Baytown Refinery. While exact timesavings were not reported they were said to be positive. This project identified its primary strength to be the development of Bechtel's Procurement Tracking System (PTS) and the assurance that RFID tags can be reused without a problem.

Rohm & Haas conducted an RFID pilot study that received, identified, and tracked Honeywell smart instruments throughout their installation and loop checking processes. As the tagged instruments arrived, an inspector read the tags and answered questions related to the installment process. Benefits from the case study include: inventory shrinkage, lower rework costs, improved tracking and inventory control, and improved data integrity.

Limitations to these pilot studies include metal interference with readers, a lot of bending down to read the tags, and sun glare (making it difficult to view the reader screen). There is also a concern that bulk items may pose problems to accurate material tracking. From the latest research done in other industries, future construction tags may have the ability to be read without interference from metal and with greater read ranges that would prevent workers from bending over so much. It is also believed that costs of the tags can be driven down if smartlabels, which are also very durable, can be applied to construction applications instead of the bulkier, yet very versatile tags.

Potential Materials Tracking Applications

Potential applications for materials tracking in commercial construction include concrete placement operations and steel frame components tracking. These applications provide viable uses because they offer incremental improvements over existing methods, reduced labor costs, real-time identification and tracking, and they provide the potential for automatic billing upon receipt of materials at a jobsite.

Concrete Placement

RFID has the ability to scan items at a distance, which allows concrete trucks to be scanned while in motion as they pass through a gate. Concrete mix design and admixtures, loading time, and delivery location could also be identified as the materials are being placed into material-handling equipment. RFID can be used to track and relate concrete cylinders to truck number and placement time. RFID can also retain very important information on the attributes of the batched construction materials such as concrete mixing time, temperature, and number of drum revolutions. An RFID concrete placement system would provide critical information for invoicing, payment, and testing through the identification of concrete trucks and concrete test cylinders, ensuring proper delivery, billing, and quality control.

Steel Frame Components and Sub-assemblies

NIST (National Institute for Standards and Technology), CII FIATECH (Fully Integrated and Automated Technology), and AISC (American Institute for Steel Construction) are partnering to provide the technical infrastructure necessary to permit real-time identification and position tracking of steel frame components and subassemblies on a construction site. Objectives include eliminating interferences caused by metal or steel and to design RFID rugged enough to withstand the tough environments encountered within construction worksites. The NIST Comp-TRAK project is addressing the problem of identifying, registering and tracking discrete construction (steel frame) components and sub-assemblies on a construction site.

This project is an ongoing research study aimed to answer four questions: (1) Was the component the right one for that location?; (2) Where was it finally positioned and what was its orientation?; (3) Were there any problems associated with the component (was it incorrectly fabricated) or with its placement into the structure?; and (4) Do the appropriate managers, engineers, and planners know this information? Presently, this information is collected manually and is not efficiently communicated to all participants in the construction process. Knowledge of this information, in an on-demand format, would enhance the construction industry by dramatically reducing cycle times, reducing the amount of rework, improving reliability, setting an industry standard, and setting the foundation for real-time 4D CAD construction project management software.

Other potential materials management operations include material takeoff, requisition, issuance, installation, inspection, shipment and export, receipt of materials, instorage maintenance of materials, and return of materials.

Tools and Equipment

The machine tool industry has been using embedded systems such as RFID technology to track tool usage and to prevent tool mishandling and wrongful installation. Although the construction industry is generally not as high tech as the manufacturing industry, tools and equipment cost can be critical to complete projects within estimated budgets. Interest in tool-tracking technologies is on the upswing because it holds the potential for reduced expenditures and tighter job costing.

Tracking

Based on Tool Watch market research conducted in 1997-98, a mechanical contracting company with 30 employees spends $30,000 annually in tool expenditures without a tracking system. However, companies with a tool tracking system had annual tool expenditures 40% lower than those companies without. This technology may prove beneficial to the commercial construction industry as well as with the tracking of high-value assets that tend to walk off of construction jobsites.

Smart gang boxes equipped with a record of RFID tagged tools inside has the potential to increase worker productivity as a result of being able to accurately and quickly identify the location of particular tools. Project tool costs could also be precisely charged to proper job number and contractors could significantly improve cost coding and inventory control of high-value assets. Other information that RFID has the capacity to provide is tool work history, tool name, and serial number. The tool's possession could also be tracked with RFID tagged workers, reducing the risk of lost or stolen tools.

HOUNDware Corporation has been using bar codes for tools tracking for quite some time, but recently they have been testing RFID tool-tracking systems with wireless data collection software.

Commercial construction is also an industry highly dependent on its equipment and operators. Many contractors do not have the capital to fully supply each jobsite with all the equipment it needs, therefore, equipment must be able to be located at all times on all projects that the contractor is constructing.

Commonly operated equipment by commercial construction contractors includes generators, air compressors, backhoes, material fork trucks, dozers, scrapers, loaders, and crawler and rubber-tired cranes. Tracking these pieces of equipment would enhance a contractor's ability to evaluate their active purpose and to better manage efficient usage of equipment on simultaneous projects. Active RFID systems would also store equipment compliance records and information regarding its operational capacities to prevent dangerous usage. In-cab readers and active tags could be powered by the power source of the equipment, minimizing the cost of the readers and tags, and these devices would have a life cycle as long as the equipment's. The GPS monitor in figure illustrates the potential to combine RFID and GPS technologies to in-cab applications. However, lack of standardization may play an important role in implementing this technology in an industry that has such a complex list of equipment, tool, and materials suppliers.

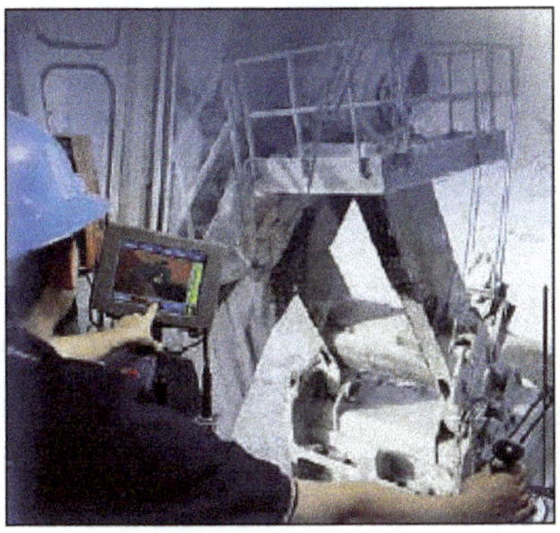

ProVision.

GPS Excavation.

Guiding

Automatic-guided-vehicle (AGV) control allows automated operations in fixed locations. RFID tags along fixed travel paths can be used to guide equipment more efficiently and effectively. The tags can simultaneously communicate storage and transfer location information with units en route. This RFID system application may be beneficial to automated construction processes of the future.

Commercial construction applications may include equipment with RFID technology that allows automated trenching and soil excavations with backhoes, automated hoe rams for rock fracturing, and instructional guidance to crane operators about load capacity and range of movement. The technology would also record the number of movements and engine run-time to improve cycle times, to track an accurate account of material quantities moved, and to create time savings for contractors.

Other uses of automatic-guided-vehicles and equipment may include paint trucks that mark roadway lanes, asphalt-paving machines, and underground bore machines. Lane marking could be much more efficient if RFID was embedded into the pavement and paint trucks with in-cab readers. This would reduce the labor costs, and these tags could be embedded during the construction of the roadway to measure asphalt pavement maturity (this will be discussed later in the report). Also, tunneling timesavings could be achieved if underground bore machines had an RFID guidance track that ensured accurate and precise movement of the equipment.

Risks and challenges for use of RFID in AGV include operator judgment decisionmaking issues. For example, in a simple trench excavation one may encounter existing utilities, under these circumstances a man-guided piece of equipment would stop operating but AGV operations would never understand the circumstances and continue excavating.

Security

RFID can also provide security to construction jobsites. Workers, operators, and equipment tagged with RFID can record and make certain proper usage and handling of equipment, materials, and documents. These systems would also ensure that only qualified equipment operators have the ability to operate restricted equipment, reducing the likelihood of misuse and accidents. Besides tracking objects and people within the jobsite, it would also secure the site from unauthorized people and vehicles.

Service and Maintenance

RFID technology can be a helpful record-keeping tool for high-value assets and equipment service logs. Applications would allow mechanics in the service bay to read and write to the tags, reducing the amount of paperwork related to warranties and time consuming maintenance logs. It will also encourage operators and managers to keep better track of their assets because there are not hundreds of different pages mixed around for a variety of tools and equipment. The equipment would display its records upon request via an in-cab smart tag display powered by the equipment. The tag would store information regarding maintenance and moderations, tire pressure and dimension for rubber-tired tools and equipment, service history, inspections, and operating or erection information and checklists.

These devices could also enhance fleet management. RFID linked to on board sensors that monitor equipment performance could determine if certain set point values are exceeded. If point values are exceeded the fleet management software would indicate a potential problem with the specific piece of equipment, prompting a mechanic to investigate the problem.

Other benefits may include quicker scaffolding erection and more accurate measuring and testing instruments, as a result of improved maintenance and calibration records. Construction managers would also see increased overall life cycles of high-value assets (tool and equipment) that contractors depend on to get projects completed on time and under budget.

Document Control

The construction industry is an industry that is very dependent upon paper for transmittals of shop drawings, plans and specifications, change orders and billing, and RFIs (requests for information). Although the Internet has allowed the use of e-mailing documentation there is still a mailed copy sent to other players involved (contractors, architects, engineers, owners, etc.). When the costs of RFID technology is more reasonable, applications that embed tags into construction documents, files, or file folders would significantly reduce the amount of time and money spent managing files, and until the industry is paperless there will always be a way to more efficiently manage documents.

Tiny integrated circuits (ICs) can be made into labels or embedded into the paper itself. The Hitachi embedded RFID tag is 0.4-mm square and is wireless accessible at 2.45 GHz allowing files to be secured and managed more efficiently. Documents equipped with RFID tags enable files to encode file identification numbers making them easier to locate and secures them from being removed without authorization from designated filing areas. This technology could be implemented with automated checkout capabilities with RFID tagged employees or an alarming system that identifies documents leaving a secured area.

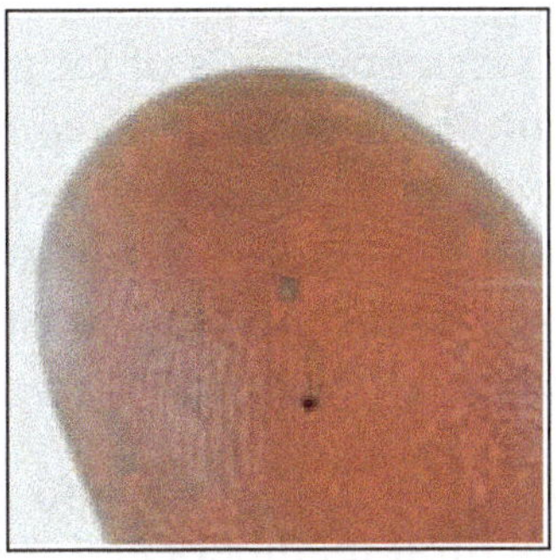

Hitachi Mu-chip.

Current Risks and challenges include the movement towards electronic file transfers and small reading ranges of the technology. The Hitachi chip is stated to have a read range of one-foot, indicating that readers would have to be installed in each file cabinet drawer and within close proximity to entrances to implement such technologies.

Service and Maintenance of Built Construction Projects

Equipment and materials with RFID systems installed in them provide benefits to contractors and owners. Contractors benefit during the construction process in several ways: accurate and quick identification and tracking of jobsite equipment and materials that are permanent fixtures in the completed project, lower costs of rework because of more efficient tracking, and automatic billing to owners, enhancing their request for payment invoicing. Owners receive the benefit of having a real-time, accurate account of materials on-site, tracking of materials through installation, and measures to evaluate total quality. They also can monitor materials and equipment for routine maintenance and failure prevention.

RFID technologies have the potential to enhance the abilities of constructors and owners to evaluate percent complete of the total project by simply walking through the site with an RFID reader. Potential is stressed because at this time read ranges are inadequate and the cost of readers and tags are not economically feasible. Risks of increased costs and lack of trust in the technology also hinder the use of RFID in service and maintenance applications.

Service Records

RFID and mechanical sensor technology could benefit HVAC, fire protection, elevator systems, and other major equipment assets through service and maintenance monitoring systems. Much like construction equipment maintenance and service monitoring, owners could more efficiently identify and manage routine maintenance and warranty information on installed parts and systems.

Failure Prevention

The use of electronics technology in civil structures, buildings, dams, bridges, and tunnels, may prevent, or at least warn engineers of fatigue or failure. Micro-size sensors will be deeply embedded within these structures, constantly supplying data on shear, strain, pressure, and other forces that can affect them. These applications would save the expense of sending out safety inspectors and engineers to monitor these structures; we will depend on computerized data centers to constantly monitor every conceivable parameter to prevent failure of civil structures.

Commercial construction may apply this technology to major structural components of buildings to strengthen structural members and to prevent their failure. A tag may be able to detect the strain in a column that is under an amount of stress equivalent to its buckling stress and provide a warning. It may also activate piezoelectric actuators, which send a strong current through the column at a specific location, creating an additional force on the column to stabilize it. Such experimentation has shown that the critical load can be increased by a factor of 2.9 before the column buckles.

Other Potential uses for RFID in Commercial Construction

Concrete and Asphalt Pavement Maturity Monitoring

RFID applications measuring temperature fluctuations within an identifiable location of placed concrete or asphalt pavement provides interesting information regarding strength and maturity.

WAKE, Inc. and IDENTEC SOLUTIONS have created an active UHF RFID tag that monitors concrete maturity by burying the tag in the concrete and taking temperature readings at user defined intervals. The raw temperature data provides the information to determine optimum concrete strength, curing rates, and documented quality control data for the project. The benefits to commercial construction contractors and the construction industry as a whole include: accelerated schedules from time savings, early formwork removal, accelerated pre-stressed release times, determination of precise sawing times for joints, identification of weak spots in the concrete, hydration rate and remedial action warning. Also, by implementing the concrete maturity method system (CMMS) of in-situ strength testing, concrete cylinders could eventually be phased out. The data from the tag is stored in the handheld PC and can be easily transferred to a desktop PC.

The temperature accuracy of the CMMS exceeds ASTM C1074-93 requirements, however, there is still an issue of trusting technology in an area of construction that ensures a safe working platform from which to build, concrete slabs and foundations. Furthermore, the cost of implementation may be an issue that the industry must decrease in order to make such RFID applications marketable.

Compaq iPac Pocket PC with transceiver.

Field Operations

RFID tags embedded in workers' hard hats can be used in the automated tracking of personnel instead of manual data recording and data entry of workers time sheets. This process would eliminate unnecessary labor expenses and provide an accurate payroll account.

Safety

RFID tags containing pertinent safety information could be attached to safety equipment such as slings, safety harnesses and belts, scaffolding, and hardhats making them easier to track and ensuring that they are in safe operating condition. The benefits of improved safety records include lower labor costs, lower direct and indirect costs associated with injuries, accidents, and deaths, and a more comfortable working environment.

Another safety application for RFID technology is hazardous materials tracking. Identifying hazardous waste and hazardous construction materials from a safe distance will prevent accidental contact and keep workers informed of the type and location of Haz-Mat on site. Underground work environments such as tunneling, mining, and work in confined spaces may find value in using RFID systems to improve safety conditions. RFID tags can be used to monitor workers, support systems protecting the workers, and equipment operating around them. In mining operations RFID has been used to prevent coal carts and conveyors from injuring workers by having shut-off signals alert the machinery when it detects a miners RFID badge.

In commercial construction, RFID systems with the capacity to warn workers that heavy equipment is operating in the immediate vicinity may prevent the number of fatalities and accidents resulting from persons coming in contact with machinery. Any number of fatalities resulting from workers caught in the blind spots of equipment and machinery is way too many. Workers and equipment armed with RFID alert systems, such as a vibrating hardhats and equipment kill switches, may prevent unnecessary accidents from happening and create a new safety awareness about the project. All too often experienced workers get too comfortable with their surrounding and new employees try to illustrate their potential through hard work are less attentive to small movements that place them in harms way. RFID systems that alert workers when equipment is nearby may have an active role in rejuvenating worker attentiveness to potential dangers.

However, like any other "new" idea, the industry cannot afford to simply implement the application, allow it to become mundane, and not follow up with another method or process to keep workers on their toes. Conversely, if tags are used to restrict equipment operators to only those that are authorized to operate the equipment, there may be a long-term safety advantage in operator security applications.

Fall protection and prevention is also a major contributor to construction accidents and fatalities. RFID applications that may enhance a worker's ability to evaluate and prevent actions leading to falls include: alerts that one is approaching a guardrail or a nearby boundary with a steep change in elevation. Guardrail alert systems would function similarly to the equipment alert examples above, however the falls from steep grade changes (major excavations) or from fatal heights (roof structures or multi-story floor elevations) may use more complex RFID technology. As in equipment guiding applications of RFID, fall protection RFID would require the layout of an RFID "danger" bound. Once the network of tags is laid-out the system would function like invisible electric dog fences, except they would be called invisible fall protection fences in construction. Challenges presented by this technology include an alert system that can effectively change the behavior of a worker and prevent him/her from mistaking courage with stupidity or ignorance. Time and costs associated with setting up such a network must also be overcome. Users must also overcome the risk of fully trusting such lifesaver systems because radio frequency reception and transmission may be influenced by interferences from other objects or frequencies.

RFID IN ATTENDANCE TRACKING

Using Radio-Frequency Identification (RFID) for attendance tracking offers a few advantages:

- Speed: To record attendance, all you need is to bring the RFID tag within a proximity of about an inch from the reader. The process is quick and there is no swiping or scanning involved.

- Cost: RFID tags are amazingly cheap (less than 30 cents each for the RFID key fob). The reader is also inexpensive (less than $30 each for a computer reader).
- Convenience: The tag is provided in many forms including a key fob that can be attached to a keychain. There is also an option for RFID wristbands. The reader is compact and light weight.
- Security: The RFID tag does not carry any personal information and can be easily replaced if lost.

HUMAN MICROCHIP IMPLANT

A human microchip implant is typically an identifying integrated circuit device or RFID transponder encased in silicate glass and implanted in the body of a human being. This type of subdermal implant usually contains a unique ID number that can be linked to information contained in an external database, such as personal identification, law enforcement, medical history, medications, allergies, and contact information.

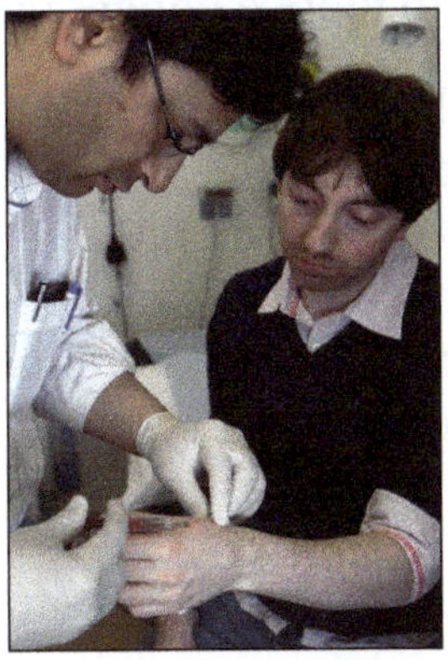

A surgeon implants British scientist Dr Mark Gasson in his left hand with an RFID microchip.

Hobbyists

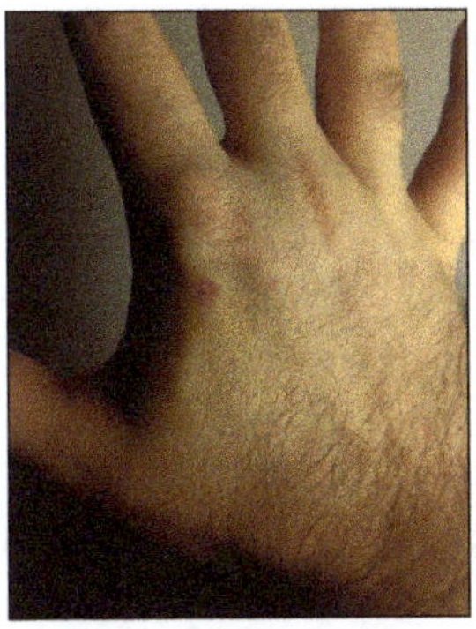

An RFID tag visible under the skin soon after being implanted.

Several hobbyists have placed RFID microchip implants into their hands or had them inserted by others.

Alejandro Hernandez CEO of Futura is known to be the first in Central America to have a Dangerous Things transponder installed in his left hand by Federico Cortes in November 2017.

Mikey Sklar had a chip implanted into his left hand and filmed the procedure. Jonathan Oxer self-implanted an RFID chip in his arm using a veterinary implantation tool.

Martijn Wismeijer, Dutch marketing manager for Bitcoin ATM manufacturer General Bytes, placed RFID chips in both of his hands to store his Bitcoin private keys and business card.

Patric Lanhed sent a "bio-payment" of one euro worth of Bitcoin using a chip embedded in his hand.

Marcel Varallo had an NXP chip coated in Bioglass 8625 inserted into his hand between his forefinger and thumb allowing him to open secure elevators and doors at work, print from secure printers, unlock his mobile phone and home, and store his digital business card for transfer to mobile phones enabled for NFC.

Biohacker Hannes Sjöblad has been experimenting with near field communication (NFC) chip implants since 2015. During his talk at Echappée Voléé 2016 in Paris, Sjöblad disclosed that he has also implanted himself between his forefinger and thumb and uses it to unlock doors, make payments, and unlock his phone (essentially replacing anything you can put in your pockets). Additionally, Sjöblad has hosted several "implant parties," where interested individuals can also be implanted with the chip.

Commercial Implants

Digital Identity

VivoKey Technologies developed the first cryptographically-secure human implantable NFC transponders in 2018. The Spark is an AES128 bit capable ISO/IEC 15693 2mm by 12mm bioglass encased injectable device. The Flex One is an implantable contactless secure element, capable of running Java Card applets (software programs) including Bitcoin wallets, PGP, OATH OTP, U2F, WebAuthn, etc. It is encapsulated in a flat, flexible 7mm x 34mm x 0.4mm flat biopolymer shell. Applets can be deployed to the Flex One before or after implantation.

Medical Records

Researchers have examined microchip implants in humans in the medical field and they indicate that there are potential benefits and risks to incorporating the device in the medical field. For example, it could be beneficial for noncompliant patients but still poses great risks for potential misuse of the device.

Destron Fearing, a subsidiary of Digital Angel, initially developed the technology for the VeriChip.

In 2004, the VeriChip implanted device and reader were classified as Class II: General controls with special controls by the FDA; that year the FDA also published a draft guidance describing the special controls required to market such devices.

About the size of a grain of rice, the device was typically implanted between the shoulder and elbow area of an individual's right arm. Once scanned at the proper frequency, the chip responded with a unique 16-digit number which could be then linked with information about the user held on a database for identity verification, medical records access and other uses. The insertion procedure was performed under local anesthetic in a physician's office.

Privacy advocates raised concerns regarding potential abuse of the chip, with some warning that adoption by governments as a compulsory identification program could lead to erosion of civil liberties, as well as identity theft if the device should be hacked. Another ethical dilemma posed by the technology, is that people with dementia could possibly benefit the most from an implanted device that contained their medical records, but issues of informed consent are the most difficult in precisely such people.

In June 2007, the American Medical Association declared that "implantable radio frequency identification (RFID) devices may help to identify patients, thereby improving the safety and efficiency of patient care, and may be used to enable secure access to patient clinical information", but in the same year, news reports linking similar devices to cancer caused in laboratory animals had a devastating impact on the company's stock price and sales.

In 2010, the company, by then called PositiveID, withdrew the product from the market due to poor sales.

In January 2012, PositiveID sold the chip assets to a company called VeriTeQ that was owned by Scott Silverman, the former CEO of Positive ID.

In 2016, JAMM Technologies acquired the chip assets from VeriTeQ; JAMM's business plan was to partner with companies selling implanted medical devices and use the RfID tags to monitor and identify the devices. JAMM Technologies is co-located in the same Plymouth, Minnesota building as Geissler Corporation with Randolph K. Geissler and Donald R. Brattain listed as its principals. The website also claims that Geissler was CEO of PositiveID Corporation, Destron Fearing Corporation, and Digital Angel Corporation.

In 2018, A Danish firm called BiChip released a new generation of microchip implant that is intended to be readable from distance and connected to Internet. The company released an update for its microchip implant to associate it with the Ripple cryptocurrency to allow payments to be made using the implanted microchip.

Building Access and Security

In February 2006, CityWatcher, Inc. of Cincinnati, OH became the first company in the world to implant microchips into their employees as part of their building access control and security system. The workers needed the implants to access the company's secure video tape room, as documented in USA Today. The project was initiated and implemented by Six Sigma Security, Inc. The VeriChip Corporation had originally marketed the implant as a way to restrict access to secure facilities such as power plants.

A major drawback for such systems is the relative ease with which the 16-digit ID number contained in a chip implant can be obtained and cloned using a hand-held device, a problem that has been demonstrated publicly by security researcher Jonathan Westhues and documented in the May 2006 issue of *Wired* magazine, among other places.

- The Baja Beach Club, a nightclub in Rotterdam, the Netherlands, once used VeriChip implants for identifying VIP guests.
- The Epicenter in Stockholm, Sweden is using RFID implants for employees to operate security doors, copiers, and pay for lunch.

Possible Future Applications

In 2017 Mike Miller, chief executive of the World Olympians Association, was widely reported as suggesting the use of such implants in athletes in an attempt to reduce problems in sport due to drug taking.

Theoretically, a GPS-enabled chip could one day make it possible for individuals to be physically located by latitude, longitude, altitude, and velocity. Such implantable GPS devices are not technically feasible at this time. However, if widely deployed at some future point, implantable GPS devices could conceivably allow authorities to locate missing persons and fugitives and those who fled from a crime scene. Critics contend, however, that the technology could lead to political repression as governments could use implants to track and persecute human rights activists, labor activists, civil dissidents, and political opponents; criminals and domestic abusers could use them to stalk and harass their victims; and child abusers could use them to locate and abduct children.

Another suggested application for a tracking implant, discussed in 2008 by the legislature of Indonesia's Irian Jaya would be to monitor the activities of persons infected with HIV, aimed at reducing their chances of infecting other people. The microchipping section was not, however, included into the final version of the provincial *HIV/AIDS Handling bylaw* passed by the legislature in December 2008. With current technology, this would not be workable anyway, since there is no implantable device on the market with GPS tracking capability.

Since modern payment methods rely upon RFID/NFC, it is thought that implantable microchips, if they were to ever become popular in use, would form a part of the cashless society. Verichip implants have already been used in nightclubs such as the Baja club for such a purpose, allowing patrons to purchase drinks with their implantable microchip.

Anti-rhetoric Claims

Cancer

In a self-published report anti-RFID advocate Katherine Albrecht, who refers to RFID devices as "spy chips", cites veterinary and toxicological studies carried out from 1996 to 2006 which found lab rodents injected with microchips as an incidental part of unrelated experiments and dogs implanted with identification microchips sometimes developed cancerous tumors at the injection site (subcutaneous sarcomas) as evidence of a human implantation risk. However, the link between foreign-body tumorigenesis in lab animals and implantation in humans has been publicly refuted as erroneous and misleading and the report's author has been criticized over the use of "provocative" language "not based in scientific fact". Notably, none of the studies cited specifically set out to investigate the cancer risk of implanted microchips and so none of the studies had a control group of animals that did not get implanted. While the issue is considered worthy of further investigation, one of the studies cited cautioned "Blind leaps from the detection of tumors to the prediction of human health risk should be avoided".

Security Risks

The Council on Ethical and Judicial Affairs (CEJA) of the American Medical Association published a report in 2007 alleging that RFID implanted chips may compromise privacy because there is no assurance that the information contained in the chip can be properly protected.

RADIO-FREQUENCY IDENTIFICATION IN SCHOOLS

Radio Frequency Identification (RFID) is a smart attendance in schools. An International School in Noida presented the Radio Frequency Identification (RFID) card, with an uncommon chip implanted for live following of children.

The utilization of RFID cards in schools helps the school management and the parents to track and screen each and every movement of their kids. RFID is small electronic devices that contain a small chip.

Distinguishing and tracking of objects can be done with this. It gives a novel identifier to the protest and the gadget must be checked to recover the recognizing data of that object.

RFID Methodology

RFID tag will be joined to the ID card and the data identified with the student will be appended to the RFID tag. At whatever point the student enters the school and swipes the ID card, the peruser at the entryway or the entryway will recognize the tag and registers the nearness of the student he minute they go before a peruser gadget at the doors.

While leaving the school premises, the student will swipe the ID card again at the door and the peruser will distinguish the tag and the framework will perceive the tag to store the information.

Benefits of RFID:

- Safety and security of students.
- Tracking of students.
- SMS alerts to parents, teachers and guardian of their children.
- Reduce bunking of class.
- Daily attendance.

These are some of the main benefits of using Radio Frequency Identification (RFID) in schools. Safety and security of students can again be ensured through best school management software. Best school information management system is a standout amongst other strategies that guarantee students wellbeing in schools.

Students, guardians, teachers and school authoritative can partake together through school management software those aides in the wellbeing of the children.

Diverse modules in best school ERP software help to manufacture a decent relationship to guardians with the school. They can track their children, guarantee a wellbeing environment and can be in home or work put with a relax and free mind without any tension.

RFID ASSET TRACKING SYSTEM

Managing and locating important assets is a key challenge for almost any business. Time spent searching for critical equipment can lead to expensive delays or down time, missed deadlines and customer commitments, and wasted labor.

Companies have traditionally addressed this problem by managing assets using serial numbers and spreadsheets, or tracking equipment with barcode labels. However, both of these methods require a significant amount of manual labor. One way to quickly improve asset management is by utilizing radio frequency identification (RFID) technology to automatically track those assets.

An RFID asset tracking system uses electromagnetic fields to transmit data from an RFID tag to a reader. Whether you are talking about office equipment in a multi-story building, wheelchairs and IV pumps in a large hospital, tools and equipment in a factory, or pumping equipment in far-flung oil fields, RFID can provide accurate, real-time tracking data for fixed and mobile assets.

With a portable RFID reader built into a mobile computer, staff can easily scan one or more asset tags without having to physically see the tags. With traditional barcode labeling, employees would have to point a barcode scanner directly at the tag in order to get an accurate scan. In some applications, that would require crawling under desks, behind racks of servers, or climbing ladders to scan hard-to-reach assets. With RFID, the reader can be several feet away and still get an accurate scan. One employee could easily scan several dozen assets in a room in just a few seconds.

RFID tags with larger on-board memory capacity can even store additional information about the asset. For applications in remote areas where connectivity to a back-end application or database might be challenging, the tag itself can store information about maintenance activities or sensor data that can be read and written to by a field technician using a mobile computer.

Barcode labels can be used for asset tracking, and do provide a more efficient alternative to manually searching for serial numbers — but a RFID asset tracking system can provide multiple advantages over barcoding. These include:

- Reading multiple tags at once, without requiring line-of-sight between the tags and scanner. One employee could inventory an entire room full of equipment in a matter of minutes.
- Critical service information can be stored on the assets themselves, enabling more accurate asset life cycle management.
- RFID tags can be integrated with sensors and GPS technology to provide asset condition data, in addition to location information.
- By integrating the solution with a wireless LAN, assets can be identified and located remotely in seconds.
- Inventory times can be reduced from days down to hours.
- Asset security can be improved by generating real-time alerts and alarms if assets are moved into unauthorized locations or taken out of a building.
- Enhanced productivity because assets can be located almost instantaneously, and most asset management tasks can be made more efficient using RFID.

Tracking large numbers of assets doesn't have to be a logistical nightmare. Automated tracking with RFID will provide a complete view of your asset fleet without the costly and time-consuming efforts involved in manually managing equipment.

Examples of RFID Asset Tracking

Assets in most all shapes and sizes can be tracked with RFID. Below are a few industries currently using RFID systems to successfully track valuable assets of all types, such as pallets, tools, and weapons.

Law Enforcement

Law enforcement personnel face a few problems regarding assets. On a daily basis, law enforcement teams interact with a wide array of assets like firearms, vehicles, IT equipment, evidence, and more. If their weapons, computers, or vehicles are misplaced or stolen, it could cost the agency thousands of dollars. If key evidence is lost, a criminal could walk free, or major lawsuits could be filed against that precinct. Loss or theft of assets can result in large fines and lawsuits, and the officer responsible could lose his or her job.

In Val Verde County, the sheriff's office is working with RFID experts to create an asset tracking solution to monitor officers' equipment in order to prevent theft or criminal activity. Using passive RFID in conjunction with real-time data-collection systems will help to ensure safety and protection of an officer's vehicle and equipment at all times.

Currently, the program remains in the pilot stage, but if it provides visibility into the movements of equipment, the program could potentially be implemented in other law enforcement precincts and set a standard for RFID asset tracking.

Manufacturing

In manufacturing, lost assets can cause major issues such as stalled work, lost sales, and wasted company money. Manufacturing companies track a variety of assets, some as small as work-related tools, to large assets like shipments of car parts. The ramifications of shipments arriving behind schedule, whether the cause be a delay in parts or loss of the assets entirely, can cost companies their reputation and clientele.

By adhering extremely rugged, orientation insensitive wet inlays on these plastic pallets, the company is able to track them among their pallet pool system between the various regions. These custom RFID tags enable the company to rent and successfully recover their reusable pallets without wasting employee time searching for them.

Education

One of the most talked about industries using RFID for asset tracking is education. From tracking elementary school children to college students, RFID has been in the news in the education industry since 2005, but what RFID is tracking now shows the strides this technology has made in the past 10 years.

RFID is being used to make schools safer for children and teachers around the nation. In a Nebraska school, RFID ensures students arrive at school safely while riding their bikes in the Ride-Your-Bike-to-School program. Solar powered RFID readers at the school read the RFID tags on each students' backpack, and that information is uploaded to the internet for parents, students, and school administrators to view.

Other school systems are tracking books (both loaned and school library books), IT equipment, and other school property enabling them to protect themselves from theft and loss. Attendance tracking is also currently deployed in colleges which will ultimately reduce time to take roll, as well as encourage student attendance.

Business Environments

All types of businesses are able to use RFID asset tracking to solve inherent problems that occur in the workplace. Common problems addressed in the business world are inefficiencies in workflow, human error, labor costs, asset loss and theft, and zero visibility into internal assets and inventory. Altogether, RFID asset tracking can improve processes and allow businesses to become a highly effective and efficient functioning business unit.

Hundreds of possible assets are available to track in the business sector ranging from small tools, to IT equipment, reusable containers, and more. RFID can help to eliminate some unnecessary manual processes that slow businesses down and keep them from operating at an optimal level.

Healthcare

With thousands of patients, employees, and assets, hospitals and healthcare offices are ideal locations to implement RFID asset tracking systems. Currently, several hospitals across the world have RFID asset tracking systems in place that utilize all types of RFID technologies including active RFID, passive RFID, RTLS, and Wi-Fi systems. Some hospitals use passive RFID to track tools in addition to other reusable assets, also they may track patients with RTLS, active RFID, or Wi-Fi based systems.

The problems hospitals face, such as lost or stolen assets, can leave patients without beds or machines critical for their treatment. Children's Hospital in Birmingham, Alabama, has started tracking assets with an RTLS system in order to manage their high-value equipment like beds, defibrillators, IV pumps, and more. The tags currently used send out IR (infrared) beacons as well as RF signals as a type of redundancy system in case one or the other is blocked by other equipment.

Using this collaborative system, the hospital is able to track, locate, and provide needed maintenance for high-value equipment with ease. Other hospitals are using the same type of system to track patients as a way to reduce wait times and increase efficiency. RTLS in collaboration with RFID provides a dynamic system which gives real time information that helps to maintain and manage different types of assets.

RFID LIBRARY MANAGEMENT SYSTEM

RFID systems use a unique numeric code stored in the memory of radio-frequency tags as an identifier. The degree of uniqueness of the code is determined by the functional needs of the automated systems in which it is used. The main requirement for the identification of code formation is its uniqueness within the boundaries of a specific system.

In the first projects of library RFID systems, a UID code of the radio-frequency label was often used as a unique identifier of accounting objects. The use of this type of identifiers in the library automation system can only ensure their uniqueness. The UID value is set during chip manufacturing; it is constant, and its structure is determined by needs of radio-frequency label manufacturers. In addition, RFID systems based on UID have significant limitations associated with existing library technology: the impossibility of identification of group accounting items in the case of application of non-inventory registration technology for documents and in the case of accounting of document sets of the book stock collections.

The use of radio-frequency labels in the RFID library system, compatible with ISO 28560 standards, involves the use of a rewritable memory area of the label to accommodate structured data, which includes data elements defined in the ISO 28560-1 standard. One of the mandatory data elements is the "Primary Item Identifier" unique for each document instance in the collection of one library. An arbitrary value, that meets the requirements of the ILS, can be assigned for this element. In this case, it is possible to identify the RFID system of group accounting items, such as the publication, as well as sets of documents. The mandatory data block is supplemented by the "Set Information" element for support of accounting of document set. The block is present by the structure of "total set/part number" elements. In addition, if document identifier is located in the rewritable memory, it becomes possible to structure it in order to support functionality of the general system by means of the RFID system.

The ISO 28560 standard defines the length of the primary item identifier as 16 bytes. If you use one byte to display a single character, you can number 10000000000 (10 quadrillion) instances of documents with direct decimal numbering. If you use alphabetic characters to form an identifier, this number of unique combinations will be much greater. Libraries with such collections of printed publications currently do not exist, and in the foreseeable future their appearance is not expected. This code space redundancy can be used to place additional information in the ID code. It can be used to extend functionality of the RFID library system. The primary item identifier must be a structure, each element of which provides a unique identification of the section of the library collection on its hierarchical level. All elements together make up the code of the primary item identifier, which must be unique within the library collection.

Entering into the primary item identifier of additional data elements is suitable for RFID systems that support automation of technological processes related to inventory, with varying degrees of autonomy from library OPAC. These data encoding method give an additional value to the identifier, and it can complement the standard method for writing data elements to the radio-frequency label memory, as defined in ISO 28560-3, 4. Such encoding methods can get faster response of the RFID system by reducing and simplifying read operations for user memory. Also it can be useful, for example, in the case of use of RFID system equipment or ready-made third-party software modules that do not fully implement functions of data elements encoding, according to ISO 28560.

It is advisable to choose data items, which are used in automation of technological operations by means of the RFID system, for encoding the primary item identifier. In addition, selected data items must also remain unchanged, because the primary item identifier must remain the same throughout the life of the document in the library collection. The "Book number" and the "Set information" data items can be used for automated verification of the book stock collections.

The book number data item indicates document location in the library store. Inserting these data item into the code structure of the primary item identifier may be efficient, if it is unchanging for document and based on unchanging classification characteristics. For example, in the case of semantic arrangement of the book stock collections, it can be compiled on the basis of library classification tables (indices UDCC, DCD, etc.) or on the basis of library collection identifier classification. In the case of formal arrangement, such features may be the book format (size and accession arrangement), document type, author number, year of publication, etc.

The book number is used in library processes, related to automated inventory, but it can also be used in other processes, for example, in document pre-ordering systems and to automatically determine possible delivery time of ordered documents from storage location to issuing location.

Entering book index into the structure of the primary item identifier is available for the collection of a separate library, since different libraries may apply different systems of collection arrangement, and formats of book index can vary.

The primary item identifier provides unique codes only within the local integrated library system. To ensure the uniqueness of several libraries, the standard defines an additional data item—"Owner Institution (ISIL)". The ISIL code is the International Standard Identifier for Libraries and related organizations. Its format is defined in ISO 15511 standard as a data structure that consists of ISO 3166-1 country code (alpha-2 type) and organization identifier as an alphanumeric element that identifies library in the national identification system.

The procedure for using the ISIL code to identify library documents is defined in the international standard ISO 20247. This standard defines the International Library Item Identifier (ILII) as a structure, consisting of two elements:

- ISIL or ISCI identifier.

- Local item identifier.

The ISCI specified as a possible element of the ILII structure is a standard collection identifier and it is defined in ISO 27730 standard. Structurally, the ISCI represents ISIL code with extension in the form of a supplementary collection identifier. Collection is defined in the standard as a logical group of one or more resources. Collections can also be logically or physically grouped or separated, i.e. a collection can be part of one or more other collections and can consist of one or more sub-collections. Collection can be an archive reading room, a digital collection of electronic resources, or OPAC of the library. Collection may consist of documents, combined on a semantic basis and located in different physical sections of the book stock store (in accordance with the type of arrangement adopted in the library) or in different sections of the virtual repository for electronic document collections. The need to use the ISCI collection identifier in the RFID system is entirely determined by the configuration of the technological system of a particular library. If you want to store it in the label data structure, the part of code that extends the ISIL code can be written to the internal code field of the Alternative Owner Institution defined as a data element in the ISO 28560-1.

In general, data structure, presented in the international standard ISO 20247, defines the method of forming identifier of the library item, which provides its unique identification on the scale of several libraries and several countries.

References

- Rfid-and-its-effect-on-supply-chain-management: fronetics.com, Retrieved 14 April, 2020
- Fister, I. Jr., Fister, I., Mernik, M., Brest, J. Design and implementation of domain-specific language Easytime. Computer Languages, Systems & Structures, 37(4), 151-167, 2011.doi:10.1016/j.cl.2011.04.001
- Attendance-tracking-rfid: engineerica.com, Retrieved 14 April, 2020
- "RFID Technology Gaining Popularity With Suppliers, Retailers". Know SCM (supply chain management). Arizona State University, W.P. Carey School of Business. 22 December 2005. Archived from the original on 4 April 2012. Retrieved 2012-12-19
- Laundry-automation, application, product: idsolutionsindia.com, Retrieved 15 May, 2020
- Raymond, Nate (2008-03-01). "Taming Texas". The American Lawyer. American Lawyer Media. Retrieved 2008-09-04

Permissions

All chapters in this book are published with permission under the Creative Commons Attribution Share Alike License or equivalent. Every chapter published in this book has been scrutinized by our experts. Their significance has been extensively debated. The topics covered herein carry significant information for a comprehensive understanding. They may even be implemented as practical applications or may be referred to as a beginning point for further studies.

We would like to thank the editorial team for lending their expertise to make the book truly unique. They have played a crucial role in the development of this book. Without their invaluable contributions this book wouldn't have been possible. They have made vital efforts to compile up to date information on the varied aspects of this subject to make this book a valuable addition to the collection of many professionals and students.

This book was conceptualized with the vision of imparting up-to-date and integrated information in this field. To ensure the same, a matchless editorial board was set up. Every individual on the board went through rigorous rounds of assessment to prove their worth. After which they invested a large part of their time researching and compiling the most relevant data for our readers.

The editorial board has been involved in producing this book since its inception. They have spent rigorous hours researching and exploring the diverse topics which have resulted in the successful publishing of this book. They have passed on their knowledge of decades through this book. To expedite this challenging task, the publisher supported the team at every step. A small team of assistant editors was also appointed to further simplify the editing procedure and attain best results for the readers.

Apart from the editorial board, the designing team has also invested a significant amount of their time in understanding the subject and creating the most relevant covers. They scrutinized every image to scout for the most suitable representation of the subject and create an appropriate cover for the book.

The publishing team has been an ardent support to the editorial, designing and production team. Their endless efforts to recruit the best for this project, has resulted in the accomplishment of this book. They are a veteran in the field of academics and their pool of knowledge is as vast as their experience in printing. Their expertise and guidance has proved useful at every step. Their uncompromising quality standards have made this book an exceptional effort. Their encouragement from time to time has been an inspiration for everyone.

The publisher and the editorial board hope that this book will prove to be a valuable piece of knowledge for students, practitioners and scholars across the globe.

Index

CPSIA information can be obtained
at www.ICGtesting.com
Printed in the USA
LVHW060907030122
707669LV00002B/388

9 781639 892327